D1256338

Introduction to Complex
Hyperbolic Spaces

Serge Lang

Introduction to Complex Hyperbolic Spaces

With 12 Illustrations

Springer-Verlag
New York Berlin Heidelberg
London Paris Tokyo

FAIRLEIGH DICKINSON UNIVERSITY LIBRARY, TEANECK, N.J.

QA
331
L2553
1987

Serge Lang
Department of Mathematics
Yale University
New Haven, CT 06520
U.S.A.

AMS Classifications: 10B15, 32C10

Library of Congress Cataloging in Publication Data
Lang, Serge
  Introduction to complex hyperbolic spaces.
  Includes index.
1. Functions of several complex variables.
2. Hyperbolic spaces. 3. Geometry, Differential.
4. Nevanlinna theory. I. Title.
QA331.L2553 1987   515.9′4   86-28037

© 1987 by Springer-Verlag New York Inc.
All rights reserved. This work may not be translated or copied in whole or in part without
the written permission of the publisher (Springer-Verlag, 175 Fifth Avenue, New York, New
York 10010, U.S.A.), except for brief excerpts in connection with reviews or scholarly
analysis. Use in connection with any form of information storage and retrieval, electronic
adaptation, computer software, or by similar or dissimilar methodology now known or
hereafter developed is forbidden.

Typeset by Composition House Ltd., Salisbury, England.
Printed and bound by R. R. Donnelley & Sons, Harrisonburg, Virginia.
Printed in the United States of America.

9 8 7 6 5 4 3 2 1

ISBN 0-387-96447-9 Springer-Verlag New York Berlin Heidelberg
ISBN 3-540-96447-9 Springer-Verlag Berlin Heidelberg New York

Foreword

Since the appearance of Kobayashi's book, there have been several results at the basic level of hyperbolic spaces, for instance Brody's theorem, and results of Green, Kiernan, Kobayashi, Noguchi, etc. which make it worthwhile to have a systematic exposition. Although of necessity I reproduce some theorems from Kobayashi, I take a different direction, with different applications in mind, so the present book does not supersede Kobayashi's.

My interest in these matters stems from their relations with diophantine geometry. Indeed, if X is a projective variety over the complex numbers, then I conjecture that X is hyperbolic if and only if X has only a finite number of rational points in every finitely generated field over the rational numbers. There are also a number of subsidiary conjectures related to this one. These conjectures are qualitative. Vojta has made quantitative conjectures by relating the Second Main Theorem of Nevanlinna theory to the theory of heights, and he has conjectured bounds on heights stemming from inequalities having to do with diophantine approximations and implying both classical and modern conjectures. Noguchi has looked at the function field case and made substantial progress, after the line started by Grauert and Grauert–Reckziegel and continued by a recent paper of Riebesehl.

The book is divided into three main parts: the basic complex analytic theory, differential geometric aspects, and Nevanlinna theory.

Several chapters of this book are logically independent of each other. The chapter on Brody's theorem can be read before hyperbolic imbeddings, for instance. The two chapters on the differential geometric aspects of the theory, Chapters IV and V, can be read separately. The last three chapters on Nevanlinna theory, Chapters VI, VII, and VIII, are essentially independent of everything that precedes them and could be used as a continuation for a standard course in complex variables. I have tried to write all these chapters so that whatever interconnection does occur between them is logically separated and can be omitted without preventing the understanding of most of the chapter. For instance, it takes Nevanlinna theory to prove that certain concretely given examples are Brody or Kobayashi hyperbolic. The examples are cited in Chapter III and proved in Chapter VII, but it is not necessary to know Chapter III or Brody's theorem in order to understand Chapter VII.

Chapter VII ends with the Second Main Theorem of Nevanlinna theory for hyperplanes in projective space $\mathbf{P}^n$. This result of 1929, due to Cartan (Nevanlinna for $n = 1$) is proved by Cartan's method which was overlooked by many people for many years. The extension to the non-linear case is a major unsolved problem, involving serious difficulties in analysis and differential geometry.

Chapter VIII, which amounts to results of Bloch and Cartan, brings to the fore results obtained sixty years ago and almost entirely forgotten until papers of Fujimoto, Green, and Kiernan–Kobayashi started setting them in the context of hyperbolicity. This chapter pushes certain analytic techniques as far as they are known to go. Again, because of Vojta's remark that taking logarithmic derivatives is the analogue of taking successive minima in number theory (see the end of [Vo]), one can expect new interest, both by the analysts and the number theorists, to be devoted to the questions considered in that chapter.

Therefore I hope this book will be useful to many types of mathematicians: analysts, differential geometers, algebraic geometers, and number theorists at the very least.

I am much indebted to Mike Artin, Daniel Horn, Kobayashi, Burt Totaro and Vojta for useful comments and corrections. I also thank Horn and Totaro for helping with the proofreading.

New Haven, Connecticut   S. LANG

Added in proofs. Walter Kaufmann-Bühler died in December 1986, while this book was in production. He was the principal person with whom I dealt at Springer-Verlag since the beginning of my association a decade ago, and the present book is the last one he handled. I trusted his judgment both in mathematical and more political editorial matters. The extent to which I miss him cannot be overestimated.

SERGE LANG

Contents

CHAPTER 0

Preliminaries 1

§1. Length Functions. 1
§2. Complex Spaces 7

CHAPTER I

Basic Properties 11

§1. The Kobayashi Semi Distance. 11
§2. Kobayashi Hyperbolic. 19
§3. Complete Hyperbolic 24
§4. Connection with Ascoli's Theorem 28

CHAPTER II

Hyperbolic Imbeddings 31

§1. Definition by Equivalent Properties. 31
§2. Kwack's Theorem (Big Picard) on **D**. 39
§3. Some Results in Measure Theory. 43
§4. Noguchi's Theorem on **D**. 55
§5. The Kiernan-Kobayashi-Kwack (K^3) Theorem and
    Noguchi's Theorem 57

CHAPTER III

Brody's Theorem 65

§1. Bounds on Radii of Discs 65
§2. Brody's Criterion for Hyperbolicity. 67
§3. Applications. 72
§4. Further Applications: Complex Tori 83

CHAPTER IV

Negative Curvature on Line Bundles   87

§1. Royden's Semi Length Function.   88
§2. Chern and Ricci Forms.   94
§3. The Ahlfors–Schwarz Lemma   102
§4. The Equidimensional Case   107
§5. Pseudo Canonical Varieties.   118

CHAPTER V

Curvature on Vector Bundles   124

§1. Connections on Vector Bundles   124
§2. Complex Hermitian Connections and Ricci Tensor   132
§3. The Ricci Function   139
§4. Garrity's Theorem   147

CHAPTER VI

Nevanlinna Theory   158

§1. The Poisson–Jensen Formula   158
§2. Nevanlinna Height and the First Main Theorem   167
§3. The Theorem on the Logarithmic Derivative.   171
§4. The Second Main Theorem   179

CHAPTER VII

Applications to Holomorphic Curves in P^n   184

§1. Borel's Theorem   185
§2. Holomorphic Curves Missing Hyperplanes   194
§3. The Height of a Map into P^n.   200
§4. The Fermat Hypersurface.   204
§5. Arbitrary Varieties   212
§6. Second Main Theorem for Hyperplanes   220

CHAPTER VIII

Normal Families of the Disc in P^n Minus Hyperplanes   224

§1. Some Criteria for Normal Families.   226
§2. The Borel Equation on **D** for Three Functions   233
§3. Estimates of Bloch–Cartan   235
§4. Cartan's Conjecture and the Case of Four Functions.   244
§5. The Case of Arbitrarily Many Functions   258

Bibliography   263
Index   269

CHAPTER 0

Preliminaries

Hermitian metrics are ubiquitous in complex analysis and differential geometry. As Grauert–Reckziegel first remarked [G–R], it is also useful to consider a function measuring length, without the triangle inequality. The first section briefly summarizes the definition and basic properties of such a function.

The second section summarizes their extension to complex spaces. We do not want to limit ourselves to complex manifolds, both for theoretical reasons, and because manifolds do not give enough flexibility in passing to subspaces and working functorially.

As a matter of notation, if H_1, H_2 are two real functions ≥ 0 on a set, we write

$$H_1 \ll H_2$$

to mean that there exists a constant C such that $H_1 \leq CH_2$.

0, §1. LENGTH FUNCTIONS

We assume readers are acquainted with hermitian metrics on a complex manifold. It turns out that one needs only part of the properties of such metrics for many applications. We now discuss this.

Suppose first that Z is a real manifold. Let E be a complex vector bundle over Z. By a **length function** on E we mean a function

$$H: E \to \mathbf{R}_{\geq 0}$$

into the real numbers ≥ 0, satisfying:

LF 1. $H(v) = 0$ if and only if $v = 0$.

LF 2. For all complex numbers $c \in \mathbf{C}$, we have

$$H(cv) = |c| H(v).$$

LF 3. H is continuous.

The first condition says that H is **positive**, meaning positive definite throughout. The second condition may be expressed by saying that H is absolutely homogeneous of degree 1. The third condition is much weaker than the usual smoothness condition on hermitian metrics. Of course, every hermitian metric gives rise to a norm which is a length function, but not conversely. If, however, H^2 is C^∞ and E is a line bundle, then one verifies easily that H is a hermitian metric in the usual sense. We shall denote a length function also by the usual absolute value sign

$$H(v) = |v|_H = |v|,$$

omitting the H if the reference to H is clear.

By a **semi length function**, we mean a function $H: E \to \mathbf{R}_{\geq 0}$ satisfying **LF 2**, and upper semi continuous. Thus we drop the positive definiteness of **LF 1**, and we weaken the continuity of **LF 3**. This definition is adjusted to the applications we have in mind. **Upper semi continuous** means: given $v \in E$ and $\varepsilon > 0$, there exists a neighborhood W of v in E such that for all $w \in W$ we have

$$H(w) < H(v) + \varepsilon.$$

Let K be a compact subset of Z and let H_1, H_2 be length functions. Then $H_1 \ll H_2$ over K.

As already mentioned in the introduction, our assertion means that there exists a constant $C > 0$ such that

$$H_1(v) \leq C H_2(v) \qquad \text{for all} \quad v \in E_x \quad \text{and all} \quad x \in K.$$

This is immediate.

Let H_1, H_2 be two length functions on E. Then $\inf(H_1, H_2)$ is a length function.

This is obvious. Note that even if H_1, H_2 are the length functions associated with hermitian metrics, then their inf is not so associated. Thus we get more flexibility in working with length functions than with hermitian metrics.

At the beginning of Chapter I, the reader will find the most basic example of a length function which we shall encounter, aside from the usual euclidean length function.

Next we come to look at distances. By a **distance** d (rather than a metric, as in metric space) on a set we mean a symmetric function

$$(x, y) \mapsto d(x, y)$$

such that $d(x, y) \geq 0$, $d(x, y) > 0$ for $x \neq y$, and d satisfies the triangle inequality. If we have only $d(x, y) \geq 0$, and the triangle inequality, then we speak of a **semi distance**.

(I use distance rather than "metric" because "metric" is also used for "hermitian metrics", for instance. Also I use the prefix "semi" which has been used previously for seminorms in a similar way.)

Given a semi distance d we can define the **open ball** $\mathbf{B}(x, s)$ centered at a point x, and of radius $s > 0$, to be the set of all points y such that $d(x, y) < s$. Similarly, we define the **closed ball** $\bar{\mathbf{B}}(x, s)$, and the **sphere** $\mathbf{S}(x, s)$ which is the set of points at distance exactly s from x. If we need to specify d in the notation for clarity in a given context, we write

$$\mathbf{B}_d(x, s), \qquad \bar{\mathbf{B}}_d(x, s), \qquad \mathbf{S}_d(x, s)$$

for the **open**, resp. **closed ball** and for the **sphere**.

The **open unit disc** in $\mathbf{C}$ will be denoted by $\mathbf{D}$. The open disc of radius r (always centered at the origin) in $\mathbf{C}$ will be denoted by $\mathbf{D}_r$.

We shall now study distances in connection with length functions.

Suppose that we are given a length function on the tangent bundle TZ. We shall then also say that it is a **length function on Z**. Then we may define the **length** of C^1 curves as follows. If $\gamma: [a, b] \to Z$ is such a curve, then we define

$$L_H(\gamma) = \int_a^b H(\gamma'(t)) \, dt = \int_a^b |\gamma'(t)| \, dt.$$

We also write L instead of L_H if the reference to the length function is clear. Instead of $\gamma'(t)$ we could also write

$$\gamma'(t) = d\gamma(t) \cdot 1$$

where $1 \in \mathbf{R}$ is viewed as a real tangent vector, with $\mathbf{R}$ identified as the tangent space of each point of the interval. If one prefers to identify a tangent vector with a derivation on the algebra of functions, then one can also write

$$\gamma'(t) = \gamma_*(d/dt).$$

By a **path** between x and y we shall mean a finite sequence of C^1 curves, such that the end point of one curve is the beginning point of the next curve. The length of a path with respect to the given length function is defined as the sum of the lengths of its component C^1 curves.

Given a length function on the tangent bundle, we may then define the corresponding **distance function** to be

$$d_H(x, y) = \inf L_H(\gamma),$$

where the inf is taken over all paths joining x to y. A priori, d_H is only a semi distance, that is d_H is ≥ 0, and satisfies the triangle inequality. We shall see in a moment that it is a distance, i.e. $d_H(x, y) \neq 0$ for $x \neq y$. If H is only a semi length function, then d_H is only a semi distance, namely d_H need not be definite.

Actually, instead of taking the inf over all paths, it suffices to take the inf over all C^1 curves between x and y. This is because a path can always be smoothed out to a C^1 curve having nearly the same length. We illustrate this on the following figure.

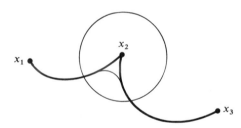

Suppose we have two curves, from x_1 to x_2, and x_2 to x_3. Take a small ball around x_2. Then the lengths of the curves inside this ball are small. Using a function like e^{-1/t^2} one knows how to construct functions whose graphs look like this:

Using such functions, we can then change the curves inside the disc so that they join smoothly, as shown on the figure.

Theorem 1.1. *If H is a length function then d_H is a distance, i.e. for $x \neq y$ we have $d_H(x, y) > 0$.*

Proof. Let $x \neq y$ be distinct points. Pick some neighborhood of y in some chart at y, represented by a ball **B** centered at y, with boundary a sphere **S**, with respect to some fixed distance function ρ which defines the topology, and such that x lies outside the closed ball $\bar{\mathbf{B}}$. Let $\gamma: [a, b] \to Z$ be a path joining x with y. Then $\rho(y, \gamma(t))$ is a continuous function of t, and so there is some point t_0 such that $\gamma(t_0)$ lies on the sphere **S**. We pick t_0 furthest to the right, so that the interval $[t_0, b]$ is mapped entirely inside the ball. If $d_H(x, y) = 0$ then we can find paths of arbitrary small lengths.

In particular, the d_H-distance between y and some point on **S** can be made arbitrarily small, using paths which lie entirely in the close ball $\bar{\mathbf{B}}$, contained in a chart. This reduces the assertion to the case of a closed ball in euclidean space, but with some length function H which may not be the euclidean length function. However, on a compact set, any two length functions are equivalent, so the corresponding distances are equivalent. This reduces the problem to the euclidean distance and two points x, y in euclidean space. Say $y = (0, \dots, 0)$ and $x = (s, 0, \dots, 0)$ for some number $s > 0$ with respect to a choice of orthonormal basis. Then the length of a curve between x and y is given by

$$\int_a^b \sqrt{x_1'(t)^2 + \cdots + x_n'(t)^2} \, dt \geqq \int_a^b |x_1'(t)| \, dt \geqq s.$$

This shows that the straight line segment in euclidean space is the shortest path between two points, and in particular the distance is $\neq 0$.

Theorem 1.2. *Let H be a length function. Then the distance d_H defines the topology of Z.*

Proof. Let ρ be a distance on Z defining the topology. We have to prove that ρ and d_H gives rise to the same notion of limit. First, if $\{x_n\}$ is a sequence in Z and $x \in Z$ are such that $\rho(x_n, x) \to 0$ then $d_H(x_n, x) \to 0$ since locally two length functions are equivalent to the same length function, say the euclidean length function in a chart. Conversely, suppose $d_H(x_n, x) \to 0$. If $\rho(x_n, x)$ does not approach 0, then there exists some $s > 0$ such that for some subsequence (which we still denote by $\{x_n\}$) each x_n lies outside the ρ-ball of radius s centered at x. As in Theorem 1.1, taking paths between x and x_n we can find a point y_n on the ρ-sphere of radius s centered at x such that

$$d_H(y_n, x) \leqq d_H(x_n, x) \to 0.$$

By local compactness, the sequence $\{y_n\}$ has a subsequence which converges to a point y on this sphere. Since d_H is continuous for the topology on Z, we get

$$d_H(y, x) = \lim d_H(y_n, x) = 0,$$

contradicting the fact that d_H is a distance. This concludes the proof.

Next we look at smooth maps and their effect on length functions.

Let $f: Y \to Z$ be a smooth map of manifolds. Then f induces the tangent map

$$Tf = f_*: TY \to TZ,$$

and we can pull back the length function by the formula

$$(f^*H)(v) = H(f_* v) \qquad \text{for} \quad v \in TY.$$

The pull-back of a length function on TZ is a semilength function on TY. The pull-back has the same degree of smoothness as the given length function, and in particular, as we have assumed H to be continuous, its pull-back is also continuous.

Instead of Tf or f_* we also write df. *Suppose we are given length functions on TY and TZ.* We can then define the **norm of $df(y)$ with respect to these length functions** by

$$|df(y)| = \sup_v |df(y)v|/|v| \qquad \text{for } v \in T_y Y, v \neq 0.$$

If H_Y, H_Z denote the length functions on TY and TZ respectively, then

$$|df(y)| \leq 1 \text{ for all } y \in Y \text{ means that } f^*H_Z \leq H_Y.$$

Suppose now that U is open in $\mathbf{C}$. Then $\mathbf{C}$ may be identified with the tangent space $T_z U$ for each $z \in U$. We can view the element $1 \in \mathbf{C}$ as a tangent vector at U. Let

$$f: U \to Z$$

be holomorphic. We adopt the notation, and **define** $f'(z)$ by

$$f'(z) = df(z) \cdot 1.$$

In other words, $f'(z)$ is the image of the standard vector 1 under the linear map $df(z)$. If tangent vectors are identified with derivations on the ring of holomorphic functions locally, then in the literature, it is also written in the form

$$f'(z) = f_*(\partial/\partial z).$$

0, §2. COMPLEX SPACES

Let Z be a complex manifold. By a **closed complex subspace** X we mean a closed subset which can be locally defined by a finite number of analytic equations. That is, given $x_0 \in X$ there exists an open neighborhood V of x in Z and a finite number of holomorphic functions $\varphi_1, \ldots, \varphi_m$ on V such that $X \cap V$ is the set of points $x \in V$ satisfying

$$\varphi_1(x) = 0, \ldots, \varphi_m(x) = 0.$$

The restriction of the holomorphic functions from V to $V \cap X$ for all choices of V open in Z defines a sheaf of functions on X, called the **sheaf of holomorphic functions** on X.

By a **complex subspace** of X we mean a locally closed complex subspace, that is a closed complex subspace of an open subset of X. Sometimes we omit the word complex, and speak just of a subspace, or also an **analytic subspace**. So the word analytic will always refer to complex analytic unless otherwise specified.

I don't want to go through a systematic treatment of the foundations of complex spaces, for which I refer to Gunning–Rossi [Gu–R]. One can define the sheaf of holomorphic functions a priori, and a map

$$f: X \to Y$$

from one complex space to another is **holomorphic** if and only if, for every holomorphic function g on an open subset V of Y, the composite $g \circ f$ (defined on $f^{-1}(V)$) is a holomorphic function on $f^{-1}(V)$.

We denote by

$$\mathrm{Hol}(X, Y)$$

the set of holomorphic maps of X into Y. When it becomes relevant, we give it the compact-open topology.

In [Gu–R] the reader will find a proof of the basic result:

Let $\{f_n\}$ be a sequence of holomorphic maps $f_n: Y \to X$ from one complex space into another. If $\{f_n\}$ converges uniformly, then the limit function is holomorphic.

In the case of complex manifolds, this is standard from courses on complex variables. The generalization to complex spaces amounts to extending $\{f_n\}$ locally in the neighborhood of every point to a uniformly convergent sequence of holomorphic maps on a complex manifold. See [Gu–R], V B, Theorem 5.

We denote by X_{reg} the open subset of X consisting of the **regular points**, i.e. the set of points such that in a neighborhood, X is a complex manifold. The complement of X_{reg} is denoted by X_{sin}, and is called the set of **singular points**. One can associate a (complex) dimension to irreducible components of a complex space locally, and it is a fact that the set of singular points in an irreducible component of dimension n has dimension $\leq n - 1$.

On a few occasions, we shall use the resolution of singularities due to Hironaka. Let X be a complex space. By a **resolution** of X we mean a holomorphic map

$$f: Y \to X$$

satisfying the following properties:

RES 1. Y is a complex manifold (not necessarily connected).

RES 2. f is proper, surjective, and a holomorphic isomorphism outside X_{sin}.

RES 3. Each point $x \in X$ has an open neighborhood U such that each component of $f^{-1}(U)$ can be imbedded as a closed subspace of $U \times \mathbf{P}^N$ for some N (commuting with the projection on U).

Hironaka's theorem asserts that for a complex space, there exists a resolution locally, that is each point has an open neighborhood which has a resolution. The Global Resolution Theorem has so far been proved only for algebraic varieties. One can construct complex spaces with ever-worsening singularities, unlike the algebraic case. In certain applications below, I prefer to use the resolution of singularities rather than other techniques, like the Weierstrass Preparation Theorem, projections on affine space, etc., which may be viewed as more "elementary". I like the formal simplicity of resolution.

We shall need length functions on a complex space. There are several ways of dealing with this. The shortest is to assume that the space, say Y, is embedded as a closed subspace of a complex manifold Z. We then view holomorphic maps of a complex space $\pi: X \to Y$ as a holomorphic map into Z. We shall need to take the norm of derivatives in the context of a holomorphic map

$$f: \mathbf{D} \to Y.$$

Then $\pi \circ f$ maps $\mathbf{D}$ into Z and we can estimate derivatives of $\pi \circ f$ as described in §1, since $\pi \circ f$ is a map from one manifold into another.

In the applications we have in mind, it does not matter whether we establish an equivalence relation between such imbeddings and length functions on the ambient manifold Z, to take into account that only their restriction to Y matters. If we want to minimize foundational questions, we then define a **length function** on Y as a pair $(Y \subset Z, H)$, consisting of a closed imbedding of Y in a complex manifold Z, and a length function H on Z. We often write (Y, H) instead of $(Y \subset Z, H)$.

On the other hand, we can localize this notion and use charts. We consider pairs $(U_i \subset Z_i, H_i)$ where $U_i \subset Z_i$ is a closed imbedding of an open subset of Y into a complex manifold Z_i, and H_i is a length function on Z_i. We say that two such pairs $(U_i \subset Z_i, H_i)$ and $(U_j \subset Z_j, H_j)$ are **compatible** if there exists a holomorphic isomorphism $\varphi: Z_i \to Z_j$ commuting with the imbeddings and such that $\varphi^* H_j = H_i$. This hypothesis implies that for any C^1 map

$$f: M \to U_i \cap U_j$$

of a manifold into $U_i \cap U_j$ we have

$$f^* H_i = f^* H_j \quad \text{on} \quad f^{-1}(U_i \cap U_j),$$

or in other words, for all $v \in T_x M$ we have

$$|df(x)v|_i = |df(x)v|_j \quad \text{if} \quad f(x) \in U_i \cap U_j \subset Z_i \cap Z_j.$$

An **atlas for a length function** on Y is then a family of compatible pairs $(U_i \subset Z_i, H_i)$ which covers Y. A maximal family of such pairs defines the **length function** itself. This allows us to define the norm $|df(x)v|$ as in §1.

The above ways are equivalent in practice, since every complex space which occurs in practice (or at least my practice) can be imbedded as a closed subspace of a complex manifold, and one has the basic fact:

Uniqueness of local imbeddings. *Locally in the neighborhood of a point of a complex space, two imbeddings into $\mathbf{C}^n$ are locally holomorphically isomorphic. In other words, given two imbeddings $\varphi_1: U \to Z_1$ and $\varphi_2: U \to Z_2$ where Z_1, Z_2 are open in $\mathbf{C}^n$, then after shrinking U, Z_1, Z_2 if necessary, there is an isomorphism $\varphi: Z_1 \to Z_2$ making the following diagram commutative:*

$$\begin{array}{ccc} & & Z_1 \\ & {}^{\varphi_1}\nearrow & \\ U & & \downarrow \varphi \\ & {}_{\varphi_2}\searrow & \\ & & Z_2 \end{array}$$

If the imbeddings are given into $\mathbf{C}^n$ and $\mathbf{C}^m$ with say $n > m$, then we can apply the basic fact to $\mathbf{C}^n$ in both cases, after taking the product of $\mathbf{C}^m$ with $\mathbf{C}^{n-m}$. The definition of compatible pairs should actually have allowed for taking such products, at the cost of the extra notation.

The essential argument for the proof of the uniqueness of local imbeddings is given in Fischer [Fi], Corollary 2, p. 18, 0.22.

We now discuss the distance function d_H associated with a length function on a complex space Y. A mapping

$$\gamma : [a, b] \to Y$$

is said to be C^1, or a **curve**, if its restriction locally in the neighborhood of every point is C^1, in the sense that if γ maps a neighborhood of a point into an open set U of Y, which is imbedded in a complex manifold as above, then the composite map into the complex manifold is C^1. The uniqueness of imbeddings locally shows that this property does not depend on the choice of local imbeddings.

In particular, if Y is globally imbedded in Z as above, then γ is C^1 as a map into Z viewed as a real manifold.

A **piecewise C^1 path** is just a sequence of C^1 curves such that the end point of one is the beginning point of the next one. A piecewise C^1 path will just be called a **path**.

We may now define the **length** of a curve, or a path, as we did for a manifold. We can also define the **distance** between two points $x, y \in Y$ to be

$$d_H(x, y) = \inf L_H(\gamma),$$

where the inf is taken over all paths joining the two points. The results of §1 are valid in this context, specifically:

Lemma 2.1. *Let Y be a connected complex space. Let $x, y \in Y$. Then there exists a path in Y connecting these two points.*

Proof. One proves that the set of points which can be connected to x by a path is both open and closed. For this, we have to distinguish regular points from singular points, and for the latter, we may use the resolution of singularities. As we shall give this argument in detail in a slightly more complicated context in Chapter I, §1, we leave it here to the reader.

Theorem 2.2. *Let Y be a complex space and H a length function.*

(i) *Then d_H is a distance, i.e. $d_H(x, y) > 0$ for $x \neq y$.*
(ii) *The distance d_H defines the topology of Y.*

Proof. The proof is the same as for manifolds.

CHAPTER I

Basic Properties

Kobayashi defined a natural semi distance on any complex space. Instead of linking two points by a chain of real curves and taking their lengths and inf via a hermitian or Riemannian metric, he joins points by a chain of discs and takes the inf over the hyperbolic metric. He calls a complex space hyperbolic when the semi distance is a distance. We shall describe the basic properties of such spaces. The most fundamental one is that a holomorphic map is distance decreasing, and hence that a family of holomorphic maps locally is equicontinuous. This gives the possibility of applying Ascoli's theorem for families of such maps. Such an application has wide ramifications, including possible applications to problems associated with Mordell's conjecture (Faltings' theorem) and possible generalizations.

I, §1. THE KOBAYASHI SEMI DISTANCE

We let $\mathbf{D}$ denote the unit disc, with its **hyperbolic norm**, or **length function**, or **metric**. If $z \in \mathbf{D}$ and $v \in T_z \mathbf{D}$ is a tangent vector at z, which in this case can be identified with a complex number, then

$$|v|_{\mathrm{hyp},z} = \frac{|v|_{\mathrm{euc}}}{1 - |z|^2},$$

where $|v|_{\mathrm{euc}}$ is the euclidean norm on $\mathbf{C}$. Instead of $\mathbf{C}$ we may write $\mathbf{C}_z$ to specify the hyperbolic metric on $\mathbf{C}$ at the point z of $\mathbf{D}$. *Note that for $z = 0$, the hyperbolic metric is the euclidean metric.*

Similarly, for any positive number r, we let $\mathbf{D}_r$ be the disc of radius r. The **hyperbolic metric** on $\mathbf{D}_r$ is defined by

$$|v|_{\text{hyp},r,z} = \frac{r|v|_{\text{euc}}}{r^2 - |z|^2} = \frac{|v/r|_{\text{euc}}}{1 - |z/r|^2}.$$

Thus multiplication by r (dilation by r)

$$\mathbf{m}_r : \mathbf{D} \to \mathbf{D}_r$$

gives an isometric holomorphic isomorphism between $\mathbf{D}$ and $\mathbf{D}_r$. For simplicity, we omit some of the subscripts sometimes, and write for instance

$$|v|_{\text{hyp}} \qquad \text{or} \qquad |v|_r$$

to denote the hyperbolic metric on $T\mathbf{D}$, or on $T\mathbf{D}_r$. The context should always make clear which metric is used, and usually for a disc, it will be the hyperbolic metric unless otherwise specified.

Recall from elementary complex analysis that the only holomorphic automorphisms of the disc up to multiplication by $e^{i\theta}$ are the maps

$$z \mapsto \frac{z - a}{1 - \bar{a}z}, \qquad a \in \mathbf{D}.$$

We shall recall below (Schwarz–Pick lemma) the proof that the holomorphic automorphisms of $\mathbf{D}$ are also metric automorphisms. Since $\text{Aut}(\mathbf{D})$ is metrically doubly transitive on $\mathbf{D}$ (given two points z, w we can first bring z to 0 and then make a rotation which brings w to a positive real number), the distance between two points can be computed as the distance between 0 and some point on the real axis. It is an easy exercise to show that among all piecewise C^1 paths between 0 and some number s with $0 \leq s < 1$, the distance is equal to the hyperbolic length of the line segment, so

$$d_{\text{hyp}}(0, s) = \int_0^s \frac{1}{1 - t^2}\, dt = \tfrac{1}{2} \log \frac{1 + s}{1 - s}.$$

Such an explicit formula will not be needed in the sequel.

In light of the metric isomorphism between $\mathbf{D}$ and $\mathbf{D}_R$ for $R > 1$, we note that if we put $s = 1/R$ then

$$d_{\text{hyp},\mathbf{D}}(0, 1/R) = d_{\text{hyp},\mathbf{D}_R}(0, 1).$$

In other words, the distance in $\mathbf{D}$ between 0 and $1/R$ is the same as the distance in $\mathbf{D}_R$ between 0 and 1.

Under an automorphism of **D**, the line segment $[0, s]$ goes into an arc of a circle perpendicular to the boundary of the disc (i.e. perpendicular to the unit circle). This is a standard fact from complex variables.

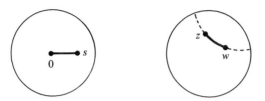

Just as in the euclidean case, the segment $[0, s]$ is the shortest path between 0 and s for the hyperbolic distance. Indeed, if $\gamma: [a, b] \to \mathbf{D}$ is any C^1 curve between these points, then

$$L_{\text{hyp}}(\gamma) = \int_a^b \frac{\sqrt{x'(t)^2 + y'(t)^2}}{1 - (x^2 + y^2)}\, dt \geqq \int_a^b \frac{|x'(t)|}{1 - x(t)^2}\, dt,$$

which proves our assertion.

The arc of circle between two points z, w in the disc is called a **geodesic** between the two points. (The circle is of course assumed perpendicular to the boundary of the disc.)

A map f between two spaces with semi distance functions d, d' is said to be **distance decreasing** if $d'(f(x_1), f(x_2)) \leqq d(x_1, x_2)$ for all pairs of points x_1, x_2 in the domain of f.

We shall investigate the behavior of the hyperbolic metric under holomorphic maps. We begin with a classical result, at the level of elementary complex variables.

Proposition 1.1 (Schwarz–Pick lemma). *Let*

$$f: \mathbf{D} \to \mathbf{D}$$

be a holomorphic map of the disc into itself. Then

$$\frac{|f'(z)|}{1 - |f(z)|^2} \leqq \frac{1}{1 - |z|^2}.$$

Proof. Fix $a \in \mathbf{D}$. Let

$$g(z) = \frac{z + a}{1 + \bar{a}z} \quad \text{and} \quad h(z) = \frac{z - f(a)}{1 - \overline{f(a)}z}.$$

Then g and h are automorphisms of the disc which map 0 on a and $f(a)$ on 0 respectively. We let

$$F = h \circ f \circ g,$$

so that $F: \mathbf{D} \to \mathbf{D}$ is holomorphic and $F(0) = 0$. Then by the chain rule,

$$F'(0) = h'(f(a))f'(a)g'(0)$$

$$= \frac{1 - |a|^2}{1 - |f(a)|^2} f'(a)$$

by a direct computation. By the ordinary Schwarz lemma, we have

$$|F'(0)| \leq 1,$$

with equality if and only if F is an automorphism, so f is an auto-morphism. We also get the reformulation

$$\frac{|f'(a)|}{1 - |f(a)|^2} \leq \frac{1}{1 - |a|^2},$$

which proves the proposition.

As already remarked in the proof of the proposition, we have equality at one point if and only if f is an automorphism. In particular, we can express the lemma invariantly in terms of the differential of f as follows. Let

$$f: \mathbf{D} \to X$$

be a holomorphic mapping into a complex hermitian manifold. Then we have an induced tangent linear map for each $z \in \mathbf{D}$:

$$df(z): T_z(\mathbf{D}) = \mathbf{C}_z \to T_{f(z)}(X).$$

Each complex space has its norm: $T_{f(z)}$ has the hermitian norm, and $T_z(\mathbf{D})$ has the hyperbolic norm. We can define the norm of the linear map $df(z)$ as usual, namely

$$|df(z)| = \sup_v |df(z)v|/|v|$$

for $v \in T_z(\mathbf{D})$, $v \neq 0$. Then the Schwarz–Pick lemma can be stated in the form:

Corollary 1.2.

   (i)   *A holomorphic map* $f: \mathbf{D} \to \mathbf{D}$ *is distance decreasing for the hyper-bolic norm.*

   (ii)   *An automorphism of* $\mathbf{D}$ *is an isometry.*

Next let X be a connected complex space. Let $x, y \in X$. We consider sequences of holomorphic maps

$$f_i: \mathbf{D} \to X, \qquad i = 1, \ldots, m$$

and points $p_i, q_i \in \mathbf{D}$ such that $f_1(p_1) = x$, $f_m(q_m) = y$, and

$$f_i(q_i) = f_{i+1}(p_{i+1}).$$

In other words, we join x to y by what we call a **Kobayashi chain of discs**. We add the hyperbolic distances between p_i and q_i, and take the inf over all such choices of f_i, p_i, q_i to define the **Kobayashi semi distance**

$$d_{\mathrm{Kob}, X}(x, y) = d_X(x, y) = \inf \sum_{i=1}^{m} d_{\mathrm{hyp}}(p_i, q_i).$$

Then d_X satisfies the properties of a distance, except that $d_X(x, y)$ may be 0 if $x \neq y$, so we call d_X a semi distance, that is d_X is ≥ 0, symmetric, and satisfies the triangle inequality.

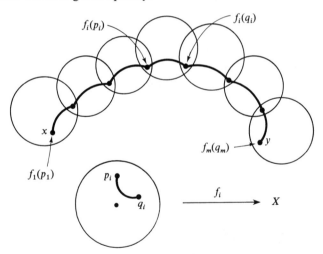

We shall call a sum $\sum d_{\mathrm{hyp}}(p_i, q_i)$ a **Kobayashi sum** (similar to a Riemann sum), and the path obtained by taking the images of the geodesics from p_i to q_i a **Kobayashi path**. The Kobayashi sum will also be called the **length** of the path (**Kobayashi length** if precision is needed).

If X is not connected, we define $d_X(x, y) = \infty$ for x, y in different connected components. Implicit in our definition of the semi distance is the following fact:

If X is connected, then there exists a chain of discs in X joining x to y, so $d_X(x, y)$ is finite.

Proof. Given x, let Z be the subset of X which can be connected to x by a chain of discs. We shall prove that Z is open and closed. Suppose $z \in Z$. There is a neighborhood U of z and a proper surjective holomorphic map $\pi: M \to U$ of a (not necessarily connected) complex manifold M onto U, and M consists of a finite number of connected components, and such that π is a holomorphic isomorphism outside the set of singular points of X in U. What we are trying to prove is obvious for a connected complex manifold, and since π is surjective, it follows that Z is open. Suppose that a sequence $\{y_n\}$ in Z converges to some point z of X. Again resolve a neighborhood of z as above. Since π is proper, we can lift $\{y_n\}$ to a sequence $\{w_n\}$ in M lying in a compact subset of M, and after picking a subsequence, we may assume without loss of generality that $\{w_n\}$ converges to a point w, such that $\pi w = z$. But for n large, w_n and w can be connected by disc in M, and the image of this disc under π connects y_n and z, thus completing the proof of our assertion.

Example 1. If $X = \mathbf{D}$ is the hyperbolic disc, then $d_\mathbf{D}$ coincides with the hyperbolic distance by Corollary 1.2 (the fact that a holomorphic map of $\mathbf{D}$ into itself is distance decreasing).

Example 2. Let $X = \mathbf{C}$ with the euclidean metric. Then $d_X(x, y) = 0$ for all $x, y \in \mathbf{C}$. Indeed, given $x \neq y$ there exists a disc of arbitrarily large radius imbedded in $\mathbf{C}$ such that 0 maps on x and $y - x$ maps to y. By a dilation, we can map $f: \mathbf{D} \to \mathbf{C}$ such that $f(0) = x$ and $f(q) = y$ where q is very close to 0. But the hyperbolic metric is very close to the euclidean metric near 0, so $d_X(x, y)$ is arbitrarily small, so equal to 0.

Directly from the definition of the Kobayashi semi distance, we conclude:

1.3 *Let $f: X \to Y$ be a holomorphic map of complex spaces. Then f is distance decreasing for the Kobayashi semi distances, that is for $x, x' \in X$ we have*

$$d_Y\big(f(x), f(x')\big) \leqq d_X(x, x').$$

Furthermore, d_X is the largest semi distance on X such that every holomorphic map $f: \mathbf{D} \to X$ is distance decreasing.

Observe that f could be an injective map of complex spaces. Thus we see (trivially) that if X is a complex subspace of Y and $x, y \in X$ then

$$d_Y(x, y) \leqq d_X(x, y), \qquad \text{that is} \quad d_Y \leqq d_X \quad \text{on } X.$$

As a consequence of the distance decreasing property, we get:

Let $f: \mathbf{C} \to X$ be holomorphic. Then for all $x, y \in f(\mathbf{C})$ we have

$$d_X(x, y) = 0$$

More generally:

1.4. If Y is a complex space such that $d_Y = 0$ and $f: Y \to X$ is holomorphic, then for all $x, x' \in f(Y)$ we have $d_X(x, x') = 0$.

The following property will be left as exercise to the reader:

1.5. Products. Let X, Y be complex spaces. For $x, x' \in X$ and $y, y' \in Y$ we have

$$d_X(x, x') + d_Y(y, y') \geqq d_{X \times Y}((x, y), (x', y'))$$
$$\geqq \max[d_X(x, x'), d_Y(y, y')].$$

1.6. Polydiscs. Let $P = \mathbf{D} \times \cdots \times \mathbf{D} = \mathbf{D}^k$ be the polydisc. Then

$$d_P((x_1, \ldots, x_k), (y_1, \ldots, y_k)) = \max_i d_{\mathbf{D}}(x_i, y_i).$$

Proof. [Ko 4], p. 47. We give the proof for two factors, $P = \mathbf{D} \times \mathbf{D}$. Since $\mathbf{D}$ is homogeneous for $\mathrm{Aut}(\mathbf{D})$, we may assume $x_1 = x_2 = 0$, and say $y_1 = x$, $y_2 = y$ with $|x| \geqq |y|$, that is

$$d_{\mathbf{D}}(0, x) \geqq d_{\mathbf{D}}(0, y).$$

Let $f: \mathbf{D} \to \mathbf{D} \times \mathbf{D}$ be defined by $f(z) = (z, (y/x)z)$. Since f is distance decreasing, we get

$$d_{\mathbf{D} \times \mathbf{D}}((0, 0), (x, y)) = d_{\mathbf{D} \times \mathbf{D}}(f(0), f(x)) \leqq d_{\mathbf{D}}(0, x)$$

as desired.

Proposition 1.7. *The Kobayashi semi distance is continuous.*

Proof. It will suffice to prove that if $y_n \to y$ then $d_X(y_n, y) \to 0$. Indeed, using the triangle inequality, if (x_n, y_n) converges to (x, y), and $d = d_X$, we have

$$d(x, y_n) \leqq d(x, y) + d(y, y_n),$$

and

$$d(x, y) \leq d(x, y_n) + d(y_n, y).$$

Then we can conclude that $d(x, y_n) \to d(x, y)$. Furthermore

$$d(x_n, y_n) \leq d(x_n, y) + d(y, y_n)$$

and the right-hand side converges to $d(x, y)$ by what we have just seen. Conversely,

$$d(x, y) \leq d(y, y_n) + d(y_n, x_n) + d(x_n, x),$$

which combined with the preceding inequality shows that $d(x_n, y_n)$ converges to $d(x, y)$.

In order to prove that $d(y_n, y) \to 0$ we distinguish cases.

Case 1. y is a regular point, that is X is a complex manifold in a neighborhood of y. Then we can pick a chart at y, say a ball B of some fixed radius, where y is represented by 0, and y_n lies in the ball, approaching the origin. There exists a fixed radius $r > 0$ such that for all n, we can imbed the disc $\mathbf{D}_r$ in B, with the origin of the disc at the origin of the ball, such that $\mathbf{D}_r$ is the set of all multiples cy_n with $c \in \mathbf{C}$ and $|cy_n| < r$. We just take the (complex) line segment passing through 0 and y_n, as large as possible inside the ball. Since the hyperbolic metric on $\mathbf{D}_r$ is close to the euclidean metric near 0, it follows that the Kobayashi distance between y_n and y approaches 0, as was to be shown.

Case 2. y is not a regular point. Suppose $d(y_n, y) \not\to 0$. Passing to a subsequence, we may assume $d(y_n, y) \geq c > 0$ for all n. By resolution of singularities, there exists an open neighborhood U of y and a proper surjective holomorphic map

$$\pi: M \to U$$

from a complex manifold M which is the union of a finite number of components. As y_n approaches y, y_n lies in a relatively compact neighborhood V of U, that is

$$V \subset \bar{V} \subset U,$$

where the closure $\bar{V}$ of V is compact, so $\pi^{-1}(V)$ is relatively compact in M. Then we can find a sequence of points $z_n \in M$ such that $\pi z_n = y_n$, and after passing to a subsequence if necessary, such that $\{z_n\}$ converges to a point z of M. Then $\pi z = y$. By Case 1, $d_M(z_n, z) \to 0$, and by the distance decreasing property of holomorphic maps, it follows that $d_X(y_n, y) \to 0$, thus proving the proposition.

I, §2. KOBAYASHI HYPERBOLIC

Let X be a complex space. We say that X is **Kobayashi hyperbolic** if the semi distance d_X is a distance, that is $x \neq y$ in X implies $d_X(x, y) > 0$. **Hyperbolic** will always mean Kobayashi hyperbolic. All other types of hyperbolic properties which we will encounter later will be subjected to a prefix to distinguish them.

Directly from the definition, we note that *to be hyperbolic is a biholomorphic invariant.*

2.1. Products. *If X, Y are hyperbolic, so is $X \times Y$.*

Proof. Immediate from 1.5.

2.2. Subspaces. *Let X be a complex subspace of Y; or let $f: X \to Y$ be holomorphic and injective. If Y is hyperbolic, so is X.*

Proof. A holomorphic map $\mathbf{D} \to X$ is also a holomorphic map of $\mathbf{D}$ into Y, so a chain of discs connecting two points of X is also a chain of discs in y. If $d_X(x, y) = 0$ it follows that $d_Y(x, y) = 0$ also. Thus our assertion is clear.

Examples. Discs and polydiscs are hyperbolic.

A bounded domain in $\mathbf{C}^n$ is hyperbolic, since it is an open subset of a product of polydiscs.

In some applications it is important to define the notion of hyperbolic for a more general set than a complex space. Thus let X be any subset of a complex space Z. We can define d_X on X (rather than on Z) by taking the maps f_i in a Kobayashi chain $f_i: \mathbf{D} \to X$ to lie in X, and be holomorphic as maps into Z. Then we obtain a semi distance on X. We say that X is **hyperbolic as a subset of** Z or **in** Z if this semi distance is a distance, that is if $x \neq y$ implies $d_X(x, y) \neq 0$. For simplicity we often omit the reference to Z if it is implied by the context. From now on, unless otherwise specified, we suppose that we deal with this situation.

Theorem 2.3 (Barth). *Let X be a connected complex space which is hyperbolic. Then d_X defines the topology on X.*

Proof. Since by definition an analytic space is locally compact with countable topology, it is metrizable by the standard Urysohn's metrization theorem, so there is a distance function ρ which defines the topology. We now have to prove that d_X and ρ are compatible, that is give rise to the same notion of limits.

First, if $\{x_n\}$ is a sequence in X such that $\rho(x_n, x) \to 0$, then $d_X(x_n, x) \to 0$ by Proposition 1.7.

Second, suppose conversely that $d_X(x_n, x) \to 0$. If $\rho(x_n, x)$ does not approach 0, then there exists some $s > 0$ such that for some subsequence (which we still denote $\{x_n\}$) each x_n lies outside the ρ-ball of radius s centered at x. We join x_n to x by a chain of discs as in the definition of the Kobayashi distance, so that the sum of the geodesics is small. The images of the geodesics in the discs of such a chain give rise to a piecewise C^1 (actually piecewise real analytic) path joining x_n to x. Let γ denote this path, and say γ has a parametric representation on some interval $a \leq t \leq b$. Then

$$t \to \rho(\gamma(t), x)$$

is a continuous function, and therefore there exists some t_0 such that $\rho(\gamma(t_0), x) = s$, in other words, $\gamma(t_0)$ lies on the ρ-sphere of radius s centered at x. Let $y_n = \gamma(t_0)$. Thus to each point x_n we have associated a point y_n on this sphere, and directly from the definition of the Kobayashi distance,

$$d_X(y_n, x) \leq d_X(x_n, x) \to 0.$$

By local compactness, the sequence $\{y_n\}$ has a subsequence which converges to a point y on the above sphere. By Proposition 1.7 we get

$$d_X(y, x) = \lim d_X(y_n, x) = 0.$$

This contradicts the hypothesis that X is hyperbolic, and concludes the proof.

The same argument of continuity shows:

Proposition 2.4. *Let ρ be a distance on X defining the topology. Suppose that X is not hyperbolic, that is $d_X(x, y) = 0$ and $x \neq y$. Given $0 < s < \rho(x, y)$ there exists a point z on the ρ-sphere of radius s centered at y such that $d_X(z, y) = 0$.*

Thus in some sense, the failure to be hyperbolic is "local", but in a much stronger sense, it is not, because even though a Kobayashi path from x to y has to cross the sphere, nevertheless the chain of discs is not localized at all, and images of a disc around y may look like this:

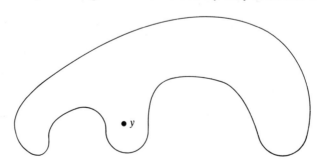

Such images may tend toward a "boundary". Hence the Kobayashi distance involves very strong global properties of the space.

Next we consider a holomorphic map, and how Kobayashi chains and sums compare under such morphisms. I have extracted a lemma which applies to several situations, and the argument in its proof stems from Kobayashi's book [Ko 4], and papers of Kiernan [Ki 2] as well as Kiernan-Kobayashi [K-K 2]. The distances d'_Y in the lemma will be taken to be Kobayashi distances, or hermitian distances, depending on the applications.

Lemma 2.5. *Let X, Y be complex spaces. Let d'_Y be a semi distance on Y, continuous for the topology of Y. Let*

$$\pi: X \to Y$$

be holomorphic and distance decreasing from d_X to d'_Y. Let $x \in X$ and $y = \pi x$. Let $\mathbf{B}(y, s)$ be the open ball with respect to d'_Y. Then there is a constant $C(s)$ depending only on s such that if we put $V = \pi^{-1}\mathbf{B}(y, 2s)$, we have for all $x' \in V$:

$$d_X(x, x') \geq \min[s, C(s)d_V(x, x')].$$

Proof. Let

$$f_i: \mathbf{D} \to X \quad (i = 1, \dots, m) \qquad \text{with} \qquad f_{i-1}(q_{i-1}) = f_i(p_i)$$

be a Kobayashi chain in X joining x to x'. Without loss of generality we may assume that $p_i = 0$ for all i. We have two cases.

Case 1. For some index j we have $\pi f_j(q_j) \notin \mathbf{B}(y, s)$. Then

$$\text{Kobayashi sum} \geq s.$$

Indeed, we have the lower estimate

$$\sum_{i=1}^{m} d_{\mathbf{D}}(0, q_i) \geq \sum_{i=1}^{m} d_X(f_i(0), f_i(q_i))$$

$$\geq \sum_{i=1}^{m} d'_Y(\pi f_i(0), \pi f_i(q_i))$$

$$\geq d'_Y(y, \pi f_j(q_j)) \geq s$$

as desired.

Case 2. For all indices i we have $\pi f_i(q_i) \in \mathbf{B}(y, s)$.

In this case, we start with a remark.

Let $f: \mathbf{D} \to Y$ be holomorphic. Let $0 < r < 1$ and let $0 < q < 1$. Then there exists a partition $[0 = t_0, t_1, \ldots, t_N]$ of the interval $[0, q]$ in $\mathbf{D}$, there are numbers r_k $(k = 1, \ldots, N)$ with $0 < r_k < r/2$, and there are automorphisms

$$g_k: \mathbf{D} \to \mathbf{D}, \qquad k = 1, \ldots, N$$

such that g_k maps $[0, r_k]$ on $[t_{k-1}, t_k]$. If we replace f by $f \circ g_1, \ldots, f \circ g_N$ then we obtain from f a Kobayashi chain joining the points $f(0)$ and $f(q)$, in other words we have a partition of the original $f: \mathbf{D} \to Y$, which has the same Kobayashi length. The advantage, as we shall see in a moment, is the inequality $0 < r_k < r/2$.

We now apply this to each f_i $(i = 1, \ldots, m)$ of the given Kobayashi chain. We let r with $0 < r < 1$ be such that

$$d_{\mathbf{D}}(0, z) < s \qquad \text{for} \quad z \in \mathbf{D}_r.$$

Thus r is a function of s only. By the above remark, without loss of generality, we may assume that our Kobayashi chain is such that $q_i < r/2$ for all i. If this new Kobayashi chain satisfies the condition of Case 1, we are done. Otherwise we are again in Case 2. Then $f_i(0) \in \pi^{-1}\mathbf{B}(y, s)$. Since $d_Y'(\pi f_i(0), \pi f_i(\mathbf{D}_r)) < s$, we get

$$f_i(\mathbf{D}_r) \subset \pi^{-1}\mathbf{B}(y, 2s) = V \qquad \text{for all } i.$$

There exists a number $C > 0$ such that

$$d_{\mathbf{D}}(0, z) > C d_{\mathbf{D}_r}(0, z) \qquad \text{for} \quad z \in \mathbf{D}_{r/2}.$$

Then the Kobayashi sum satisfies the inequalities:

$$\sum_{i=1}^{m} d_{\mathbf{D}}(0, q_i) \geq C \sum_{i=1}^{m} d_{\mathbf{D}_r}(0, q_i)$$

$$\geq C \sum_{i=1}^{m} d_{\mathbf{D}}(0, q_i/r)$$

$$\geq C \, d_V(x, x').$$

Indeed $f_i(\mathbf{D}_r) \subset V$ for all i. So if we denote by $\mathbf{m}_r$ multiplication by r, then $\{f_1 \circ \mathbf{m}_r, \ldots, f_m \circ \mathbf{m}_r\}$ is a Kobayashi chain in V. This proves Lemma 2.5.

The next proposition applies to "families" when we think of X as lying above a base space Y. In this case, we shall take $d_Y' = d_Y$ to be the Kobayashi semi distance. However, in a later application, d'_Y will be a hermitian distance. The lemma was formulated so as to be useful in many contexts.

Proposition 2.6. *Let $\pi: X \to Y$ be a holomorphic map of complex spaces. Assume that Y is hyperbolic and that to each $y \in Y$ there is neighborhood U such that $\pi^{-1}(U)$ is hyperbolic. Then X is hyperbolic.*

This proposition is a special case of a result which applies to a suitable distance function d'_Y rather than d_Y as in the lemma, namely:

Proposition 2.7. *Let X, Y be complex spaces and let d'_Y be a distance function on Y defining the topology of Y. Let $\pi: X \to Y$ be holomorphic, and assume:*

(i)   *π is distance decreasing from d_X to d'_Y.*
(ii) *To each point $y \in Y$ there is an open neighborhood U such that $\pi^{-1}(U)$ is hyperbolic.*

Then X is hyperbolic.

Proof. Let x, $x' \in X$. If $\pi x \neq \pi x'$ then the hypothesis that π is distance decreasing shows that $d_X(x, x') > 0$. So assume $\pi x = \pi x' = y$. The neighborhood U contains a ball $\mathbf{B}(y, 2s)$ for some $s > 0$ for the distance d'_Y, and $\pi^{-1}\mathbf{B}(y, 2s)$ is hyperbolic as a subspace of $\pi^{-1}(U)$. Lemma 2.5 shows that $d_X(x, x') \neq 0$. This concludes the proof.

In Proposition 2.6, of course, we use Theorem 2.3 to guarantee that d_Y defines the topology, and we take $d'_Y = d_Y$.

Remark. The assumption that $\pi^{-1}(U)$ is hyperbolic is satisfied for instance if there exists a hyperbolic space Z such that we have a complex analytic isomorphism

$$\pi^{-1}(U) \approx U \times Z.$$

Then Y hyperbolic implies U hyperbolic, implies $U \times Z$ hyperbolic. This special case when the fiber space is locally trivial holomorphically, is originally due to Kiernan, but is extremely rare. For instance a family of compact Riemann surfaces is almost never locally trivial. The proof which applied to the locally trivial case, however, applied here. Cf. [Ko 4], IV, Theorem 4.14. The same remark applies to the complete hyperbolic case discussed in the next section.

Proposition 2.8 (Coverings). *Let $\pi: X' \to X$ be a covering of complex spaces. Then X' is hyperbolic if and only if X is hyperbolic.*

Proof. If X is hyperbolic then X' is hyperbolic by Proposition 2.6. Conversely, assume that X' is hyperbolic. Let x, $y \in X$ be such that $d_X(x, y) = 0$. Let x' be a point in X' above x. Any Kobayashi chain joining x to y can be lifted to a Kobayashi chain in X' joining x' to some y' above y, with the same Kobayashi length. As these lengths tend to 0, this means that $d_{X'}(y', x')$ tends to 0. By Theorem 2.3, this implies that y' approaches x'. Therefore $\pi(y') = y$ approaches $\pi(x') = x$. Hence $x = y$, thus concluding the proof.

I, §3. COMPLETE HYPERBOLIC

We shall first reproduce [Ko 4], IV, §5. We say that a complex space X is **complete hyperbolic** if X is hyperbolic and is complete with respect to the distance d_X.

The property of being complete hyperbolic is obviously a biholomorphic invariant.

If X is hyperbolic and compact, then X is complete hyperbolic.

We now prove less obvious criteria.

Proposition 3.1. *Let X be a hyperbolic connected complex space. Then X is complete if and only if for all $x \in X$ and all $r > 0$, the closed ball $\bar{B}(x, r)$ is compact.*

Proof. If each ball is compact, it follows trivially that X is complete. We prove the converse.

Let X be a complex space and Y any subset. We define

$$U(Y, r) = \text{set of points } x \in X \text{ for which there exists } y \in Y$$
$$\text{satisfying } d_X(x, y) < r.$$

In other words, $U(Y, r)$ is the set of points in X which are at distance $< r$ from some point of Y.

Lemma 3.2. *Let X be a complex space and $a \in X$. Let $r, r' > 0$. Then*

$$U[U(a, r), r'] = U(a, r + r').$$

Proof. The inclusion $\subset$ is trivial, and uses only the triangle inequality. We prove the converse inclusion. Let $x \in U(a, r + r')$. Let ε be such that

$$d_X(a, x) < r + r' - 3\varepsilon.$$

There is a Kobayashi path $\{\gamma_1,\ldots,\gamma_m\}$ images of geodesics such that

$$d_X(a, x) \leq \text{Kobayashi sum} < d_X(a, x) + \varepsilon.$$

Let j be the largest integer such that the length of the partial path $\{\gamma_1,\ldots,\gamma_{j-1}\}$ is $< r - \varepsilon$. We subdivide γ_j into two geodesics, γ_j' and γ_j'' by some point x_j on γ_j such that the length of $\{\gamma_1,\ldots,\gamma_j'\}$ is equal to $r - \varepsilon$. Then $d(a, x_j) < r$, and from the Kobayashi path $\{\gamma_1,\ldots,\gamma_j', \gamma_j'',\ldots,\gamma_m\}$ we get

$$d_X(x_j, x) \leq d_X(a, x) + \varepsilon - (r - \varepsilon) < r + r' - 3\varepsilon + 2\varepsilon - r$$

$$\leq r' - \varepsilon.$$

This concludes the proof of the lemma.

The rest of the proof has nothing to do with complex spaces any more and can be stated as follows.

Proposition 3.3. *Let X be a locally compact space with a distance function d satisfying the equality*

$$U[U(a, r), r'] = U(a, r + r')$$

for all $a \in X$ and all positive numbers r, r'. Then X is complete for this distance function if and only if the closure $\bar{U}(x, r)$ is compact for all $x \in X$ and all positive numbers r.

Proof. If every closed ball $\bar{U}(a, r)$ is compact for all $a \in X$, then trivially X is complete. The proof of the converse depends on a lemma.

Lemma 3.4. *Let X be as in Proposition 3.3. Let $a \in X$ and $r > 0$. If there exists some $s > 0$ such that $\bar{U}(x, s)$ is compact for every $x \in U(a, r)$ than $\bar{U}(a, r)$ is compact.*

Proof. Let $t > 0$ be such that $t < r$ and $\bar{U}(a, t)$ is compact. Such t exists because X is locally compact. It will suffice to show that $\bar{U}(a, t + (s/2))$ is compact. Let $\{x_n\}$ be a sequence in $\bar{U}(a, t + (s/2))$. We must show that $\{x_n\}$ has a convergent subsequence. By assumption for each n there exists a point $y_n \in \bar{U}(a, t)$ such that

$$d_X(x_n, y_n) < \tfrac{3}{4}s.$$

After taking a subsequence if necessary, we may assume that $\{y_n\}$ converges to some $y \in \bar{U}(a, t)$. Then $\bar{U}(y, s)$ contains x_n for all n sufficiently large. Since $\bar{U}(y, s)$ is compact by assumption, the sequence $\{x_n\}$ converges to some x in $\bar{U}(y, s)$. This point x also lies in the closed set $\bar{U}(a, t + (s/2))$, thus concluding the proof.

We now finish the proof of Proposition 3.3. Assume X complete. By Lemma 3.4 it suffices to prove that there exists some $s > 0$ such that for all $x \in X$ the closed ball $\bar{U}(x, s)$ is compact. Assume the contrary. Then there exists some $x_1 \in X$ such that $\bar{U}(x_1, 1/2)$ is not compact. Again by Lemma 3.4 there exists $x_2 \in \bar{U}(x_1, 1/2)$ such that $\bar{U}(x_2, 1/2^2)$ is not compact. Continuing in this way, there exists $x_n \in \bar{U}(x_{n-1}, 1/2^{n-1})$ such that $\bar{U}(x_n, 1/2^n)$ is not compact. By assumption, the Cauchy sequence $\{x_n\}$ converges to a point x. Since X is locally compact, there exists a closed ball $\bar{U}(x, t)$ for some $t > 0$ such that $\bar{U}(x_n, 1/2^n)$ is contained in $\bar{U}(x, t)$ for sufficiently large n, and hence is compact. This contradiction proves Proposition 3.3.

Remark. Proposition 3.3 is a variant of a standard result in differential geometry due to Rinow, who recognized the generality of the result, usually phrased in terms of "inner distances", see [Ri], p. 172. Royden has observed that the Kobayashi distance is in fact an inner distance. However, I have followed Kobayashi's exposition for the proof.

Next we consider a morphism in the same context as Proposition 2.7. It is convenient to make a definition, inspired from the notion of being relatively compact. Let $\pi: X \to Y$ be a map. We say that (X, π) is **relatively complete** in Y if X is hyperbolic, and every d_X-Cauchy sequence $\{x_n\}$ in X is such that $\{\pi x_n\}$ converges to a point in Y. Note that convergence in Y is relative to the topology, and does not depend on a choice of distance function inducing this topology.

Remark. If X is hyperbolic, relatively compact in Y then (X, π) is relatively complete. Also if π is distance decreasing from d_X to a distance function d_Y' which induces the topology on Y, and if Y is complete with respect to d_Y', then (X, π) is relatively complete in Y. In practice, one or the other of these conditions is satisfied whenever one wants them.

Proposition 3.5. *Let X, Y be complex spaces. Let d_Y' be a distance function on Y defining the topology of Y. Let*

$$\pi: X \to Y$$

be holomorphic. Assume:

(i) *(X, π) is relatively complete in Y.*
(ii) *π is distance decreasing from d_X to d_Y'.*

(iii)   *To each $y \in Y$ there is a neighborhood U such that $\pi^{-1}(U)$ is complete hyperbolic.*

Then X is complete hyperbolic.

Proof. After Proposition 2.7, we have to prove that X is complete. Let $\{x_n\}$ be d_X-Cauchy in X such that $\{\pi x_n\}$ converges to $y \in Y$. Let U be as in (iii). Then $x_n \in \pi^{-1}(U)$ for all sufficiently large n. Let $s > 0$ be such that $U \supset \mathbf{B}(y, 2s)$ (ball with respect to d'_Y). For m, n large, $d_X(x_m, x_n)$ is small, and there is a Kobayashi chain in X joining these two points whose sum is small. Then by Case 1 and Case 2 of Lemma 2.5 the Kobayashi sum must be $\geq C(s)d_V(x_m, x_n)$ where $V = \pi^{-1}\mathbf{B}(y, 2s)$. Therefore $d_V(x_m, x_n) \to 0$ as $m, n \to \infty$. But the Kobayashi distance of $\pi^{-1}(U)$ is smaller than d_V because there are more Kobayashi chains in $\pi^{-1}(U)$ than in V. Hence $\{x_n\}$ converges to a point in $\pi^{-1}(U)$, thereby proving that X is complete hyperbolic, as desired.

Remark. Proposition 3.5 is applied often in two extreme cases of the map π, namely when π is surjective and is interpreted as a family of spaces $\{\pi^{-1}(y)\}$, for $y \in Y$; and also when $\pi \colon X \to Y$ is an inclusion. In the first case, we are interested in a total space X, and usually we take $d'_Y = d_Y$ to be the Kobayashi distance on Y. In the second case, we are interested in an imbedding of X in Y, and when this imbedding is a "hyperbolic imbedding", to be discussed in Chapter II, §5. Then we take d'_Y to be a distance function coming from some length function, say a hermitian distance. Proposition 3.5 provides a common ground for these two cases.

Proposition 3.6.

(a)   *Discs and polydiscs are complete hyperbolic.*
(b)   *Finite products of complete hyperbolic spaces are complete hyperbolic.*
(c)   *A closed complex subspace of a complete hyperbolic space is complete hyperbolic.*
(d)   *Let $\pi \colon X \to Y$ be a holomorphic map of complex spaces. Assume that Y is complete hyperbolic, and that to each $y \in Y$ there is a neighborhood U such that $\pi^{-1}(U)$ is complete hyperbolic. Then X is complete hyperbolic.*
(e)   *Let $\pi \colon X' \to X$ be a covering. Then X is complete hyperbolic if and only if X' is complete hyperbolic.*

Proof. The first three assertions are immediate. Part (d) is a special case of Proposition 3.5, taking $d'_Y = d_Y$.

Finally, we consider a covering as in (e). If X is complete hyperbolic, so is X' by (d). Conversely, assume X' is complete hyperbolic. We shall use the criterion of Proposition 3.1. Let $x \in X$ and let $r > 0$. Let $s > 0$.

Let $y \in \bar{\mathbf{B}}(x, r)$. There exists a Kobayashi chain in X joining x to y, and of length $< r + s$. Given x' above x in X', we can lift this chain to a Kobayashi chain in X', joining x' to some point y' above y. The lifted chain has the same length, so $y' \in \bar{\mathbf{B}}(x', r + s)$. Hence

$$\bar{B}(x, r) \subset \pi \bar{\mathbf{B}}(x', r + s),$$

and $\bar{\mathbf{B}}(x', r + s)$ is compact, so $\bar{\mathbf{B}}(x, r)$ is compact, thus concluding the proof.

Example. Let $\mathbf{D}^*$ be the punctured unit disc $\mathbf{D} - \{0\}$. Then $\mathbf{D}^*$ is complete hyperbolic because its universal covering space is the upper half plane, under the map

$$z \mapsto e^{2\pi i z}.$$

Proposition 3.7. *Let X be a complete hyperbolic complex space, and let f be a bounded holomorphic function. Then the open subset X_f consisting of those points x such that $f(x) \neq 0$ is complete hyperbolic.*

Proof. After multiplying f by a small positive number, we may assume that $f: X \to \mathbf{D}$ maps X into the unit disc. We let $U = X_f$ to avoid double subscripts. Let $\{x_n\}$ be a Cauchy sequence in U. Since $d_U \geq d_X$ on U, it follows that $\{x_n\}$ is a Cauchy sequence with respect to d_X, and so converges with respect to d_X to a point $x \in X$. We must now show $x \in U$. We have the inequalities

$$d_{\mathbf{D}}\big(f(x_n), f(x_m)\big) \leq d_{\mathbf{D}^*}\big(f(x_n), f(x_m)\big) \leq d_U(x_n, x_m) \to 0$$

as $m, n \to \infty$, so $\{f(x_n)\}$ is Cauchy in $\mathbf{D}^*$ with respect to $d_{\mathbf{D}^*}$, so convergent to a point of $\mathbf{D}^*$. Hence $\{f(x_n)\}$ converges to that same point with respect to $d_{\mathbf{D}} \leq d_{\mathbf{D}^*}$, and so $f(x) \neq 0$. Since d_U is continuous for the topology on U, it follows that $d_U(x_n, x) \to 0$ as $n \to \infty$, thereby proving the proposition.

I, §4. CONNECTION WITH ASCOLI'S THEOREM

Next we consider hyperbolicity in the context of Ascoli's theorem, which we recall.

Ascoli's theorem. *Let X be a compact subset of a metric space, and let Y be a complete metric space. Let Φ be a subset of the set of continuous maps $C(X, Y)$ with sup norm. Then Φ is relatively compact in $C(X, Y)$ if and only if the following two conditions are satisfied:*

ASC 1. *Φ is equicontinuous.*

ASC 2. *For each* $x \in X$, *the set* $\Phi(x)$ *consisting of all values* $f(x)$ *for* $f \in \Phi$ *is relatively compact.*

That Φ is **equicontinuous at a point** x_0 means that given ε, there exists δ such that whenever $x \in X$ and $d(x, x_0) < \delta$, then

$$d(f(x), f(x_0)) < \varepsilon \qquad \text{for all} \quad f \in \Phi.$$

We say that Φ is **equicontinuous on** X if Φ is equicontinuous at every point of X. As a corollary of Ascoli's theorem, one has:

Corollary. *Let* X *be a metric space whose topology has a countable base* $\{U_i\}$ *such that the closure* $\bar{U}_i$ *of each* U_i *is compact. Let* $\{f_n\}$ *be a sequence of continuous maps of* X *into a complete metric space* Y. *Assume that* $\{f_n\}$ *is equicontinuous (as a family of maps), and is such that for each* $x \in X$ *the closure of the set* $\{f_n(x)\}$ ($n = 1, 2, \ldots$) *is compact. Then there exists a subsequence which converges pointwise to a continuous function* f, *and such that the convergence is uniform on every compact subset.*

Most books on Real Analysis have a proof of Ascoli's theorem, which is standard, e.g. mine, Chapter III, §3. Note that every metric space can be metrically imbedded in a Banach space, essentially by using the distance function, see my *Real Analysis*, Chapter II, §2, p. 43. This is trivial, but useful notationally since instead of writing the distance between two points as $d(x, y)$ one can write $|x - y|$ with the norm notation.

Now suppose that ρ is a distance on X defining the topology, and that d_X is a semi distance which is continuous for the topology. Let d_Y be the given distance on Y. We apply Ascoli's theorem by using the trivial but fundamental fact:

The set of maps $f : X \to Y$ *which are distance decreasing from* d_X *to* d_Y *is equicontinuous with respect to the given topology on* X *and the given distance on* Y.

Indeed, given ε and $x_0 \in X$ there exists δ such that if $\rho(x, x_0) < \delta$ then $d_X(x, x_0) < \varepsilon$, because d_X is continuous; and if f is distance decreasing as above, then $d_Y(f(x), f(x_0)) < \varepsilon$, which proves the equicontinuity.

For brevity, we make the following definition. Let Φ be a subset of $\text{Hol}(X, Y)$. We say that Φ is **relatively locally compact** in $\text{Hol}(X, Y)$ if given a sequence $\{f_n\}$ in Φ there exists a subsequence which converges uniformly on every compact subset of X to an element of $\text{Hol}(X, Y)$. Then we get:

Theorem 4.1. *Let* X *be a complex space with Kobayashi semi distance* d_X, *and let* Y *be complete hyperbolic. Let* Φ *be a subset of* $\text{Hol}(X, Y)$

such that for each $x \in X$, the set $\Phi(x)$ is relatively compact. Then Φ is relatively locally compact in $\text{Hol}(X, Y)$.

Note that if Y is compact, then the condition on Φ is automatically satisfied.

We shall deal especially with the following special case. Suppose Y is a subspace of Z. We say that $\text{Hol}(X, Y)$ is **relatively locally compact in** $\text{Hol}(X, Z)$ if given an infinite family in $\text{Hol}(X, Y)$ there is a subsequence $\{f_n\}$ which converges uniformly on every compact subset to an element of $\text{Hol}(X, Z)$. This notion will play an important role for hyperbolic embeddings in the next chapter, and also in Chapter VIII.

CHAPTER II

Hyperbolic Imbeddings

This chapter and the next chapter on Brody's theorem are essentially logically independent. The reader interested in Brody's theorem should skip this chapter at first, and come back to it only as needed to get the extra information that under certain circumstances imbeddings are hyperbolic.

Here we go deeper into a direct study of hyperbolic behavior on the boundary of a hyperbolic space imbedded in another space. Such behavior is important in applications to families of spaces, such that the general member of the family is hyperbolic, but special members are not. One wants to know how much the Kobayashi distance can degenerate towards the boundary. It seems to be a general pattern that for "minimal models" of the families, in various contexts, the Kobayashi distance does not degenerate substantially. Formally, this is the notion of hyperbolic imbeddings. Kobayashi–Ochiai [K–O 1] prove such a result for moduli spaces, and Noguchi [No 3] does it for families of compact Riemann surfaces of genus $\geqq 2$. The problem remains open in other cases.

II, §1. DEFINITION BY EQUIVALENT PROPERTIES

In a number of applications it is important to know how the Kobayashi distance behaves at points on the boundary of a hyperbolic space. To what extent does the distance degenerate to a semi distance? In this section, we study this phenomenon, and obtain results of Kiernan [Ki 2] and Kiernan–Kobayashi [K–K 2]. We follow both.

For the rest of this section we let X be a complex subspace of a complex space Y.

We shall need to measure derivatives, so we suppose given a length function on Y. As far as we are concerned, the reader may assume that Y is a closed subspace of a complex manifold, and norms of derivatives can be measured in terms of a length function on this manifold.

The following conditions are equivalent and define what we mean for X to be **hyperbolically imbedded** in Y. We let $\bar{X}$ be the closure of X in Y.

HI 1. X is hyperbolic, and given two sequences $\{x_n\}$, $\{y_n\}$ in X converging to points x, y in $\bar{X} - X$ respectively, if $d_X(x_n, y_n) \to 0$ then $x = y$.

HI 2. Let $\{x_n\}$, $\{y_n\}$ be two sequences in X converging to points x, y in $\bar{X}$ respectively. If $d_X(x_n, y_n) \to 0$ then $x = y$.

HI 3. Given a length function H on Y, there exists a positive continuous function φ on Y such that for all holomorphic maps $f: \mathbf{D} \to X$ we have

$$f^*(\varphi H) \leqq H_{\mathbf{D}}.$$

By φH we mean of course the function such that

$$(\varphi H)(v) = \varphi(y)H(v) \qquad \text{for} \quad v \in T_y.$$

Such a function can be extended to a complex manifold in which Y is embedded, either globally or locally depending on the conventions adopted to define a length function. Recall that by positive we mean strictly positive throughout. We also remark that if X is relatively compact in Y, then the continuous function can be taken to be a positive constant since φ has positive minimum on $\bar{X}$, and only $\bar{X}$ is relevant in the inequality.

HI 4. There exists a length function H on Y such that for all holomorphic maps $f: \mathbf{D} \to X$ we have

$$f^*H \leqq H_{\mathbf{D}}.$$

Remark. *From condition* **HI 4** *it follows at once that the inclusion* $\pi: X \to Y$ *is distance decreasing from* d_X *to the distance* d_H.

Indeed, we look at Kobayashi sums, and apply the inequality of **HI 4** to each curve coming from a geodesic in the Kobayashi chain to get the asserted inequality.

HI 5. Let x, $y \in \bar{X}$ and $x \neq y$. Then there exist open neighborhoods U of x and V of y in $\bar{X}$ such that

$$d_X(U \cap X, V \cap X) > 0.$$

As usual, the distance between two sets S and T is defined as

$$d_X(S, T) = \inf d_X(u, v) \qquad \text{for } u \in S \text{ and } v \in T.$$

We shall now prove the equivalences.

Assume **HI 1**. The essential difference between **HI 1** and **HI 2** is whether x or y lies in the boundary $\bar{X} - X$. If both $x, y \in X$ then what we want comes merely from the definition of hyperbolicity. Suppose $x \in X$ and $y \in \bar{X} - X$. Suppose also $d_X(x_n, y_n) \to 0$. Since $y \notin X$, there is a ball $\bar{\mathbf{B}}(x, s)$ such that $y_n \notin \bar{\mathbf{B}}(x, s)$ for all large n, where the ball is taken with respect to d_X. Also $x_n \in \bar{\mathbf{B}}(x, s/2)$ for all large n. This is impossible, thus proving **HI 2**.

Assume **HI 2**. We shall prove **HI 3**. Let K be a compact subset of Y. We first prove that there exists a positive constant c such that for every $f: \mathbf{D} \to X$ we have

$$f^*(cH) \leq H_{\mathbf{D}} \quad \text{at every point of} \quad f^{-1}(K).$$

If not, then there exists a sequence of holomorphic maps $f_n: \mathbf{D} \to X$ and some point $z_n \in f_n^{-1}(K) \subset \mathbf{D}$ such that $|df_n(z_n)|$ is large. Since $\mathbf{D}$ is homogeneous for $\mathrm{Aut}(\mathbf{D})$, we may assume $z_n = 0$, so

$$|df_n(0)| \to \infty.$$

Passing to a subsequence if necessary, we can assume that

$$\{f_n(0)\} \quad \text{converges to some point} \quad y \in K.$$

Let U be open neighborhood of y in Y, equal to a closed subspace of $\mathbf{D}_r^m \subset \mathbf{D}_{r'}^m$ with $r < r'$. We claim

For each positive integer k there is some $z_k \in \mathbf{D}$ and an integer n_k such that $|z_k| < 1/k$ and $f_{n_k}(z_k) \notin U$.

Otherwise, there is a number $r < 1$ such that $f_n(\mathbf{D}_r) \subset U$ for all $n \geq n_0(r)$. By Ascoli's theorem, since $f_n(0) \to y$, there exists a subsequence of $\{f_n | \mathbf{D}_r\}$ which converges uniformly on every compact subset of $\mathbf{D}_r$. But this is impossible since

$$|df_n(0)| \to \infty.$$

Now let

$$y_k = f_{n_k}(0) \qquad \text{and} \qquad x_k = f_{n_k}(z_k).$$

We can pick z_k such that x_k lies in some compact set containing U. After taking a subsequence if necessary, we may assume that $\{x_k\}$ converges to a point $x \neq y$. But

$$d_X(x_k, y_k) \leqq d_{\mathbf{D}}(0, z_k) \to 0 \qquad \text{as} \quad k \to \infty.$$

This contradicts **HI 2**. The argument in fact gives us the intermediate property

HI 2$\frac{1}{2}$. *Given a compact subset K of Y, there exists $c > 0$ such that for all holomorphic $f: \mathbf{D} \to X$ with $f(0) \in K$ we have $|df(0)| \leqq c$.*

Since two length functions are equivalent on a compact set, this property is independent of the choice of length function.

Now for each compact subset K of Y there exists a constant c such that the desired inequality is satisfied. Now let $K_1 \subset K_2 \subset \cdots$ be a sequence of compact subsets of Y whose union is Y, and $K_i = \bar{U}_i$ where U_i is open, $\bar{U}_i \subset U_{i+1}$. For each K_i we find a constant c_i as above. Then we can find a positive continuous function φ on Y satisfying

$$c_i^{-1} \geqq \varphi \quad \text{on } K_i,$$

thus proving that **HI 2** $\Rightarrow$ **HI 2$\frac{1}{2}$** $\Rightarrow$ **HI 3**.

Trivially **HI 3** $\Rightarrow$ **HI 4** by using φH, which is a length function.

Assume **HI 4**. Let $x, y \in \bar{X}$ and $x \neq y$. Let $U = \mathbf{B}_H(x, s)$ and

$$V = \mathbf{B}_H(y, s)$$

be the balls of radius s with respect to the H-distance. We pick s so small that $\mathbf{B}_H(x, 2s) \cap \mathbf{B}_H(y, 2s)$ is empty. Let $x' \in \mathbf{B}_H(x, s) \cap X$ and $y' \in \mathbf{B}_H(y, s) \cap X$. By **HI 4** we know that

$$d_X(x', y') \geqq d_H(x', y') \geqq s.$$

This proves **HI 5**.

It is obvious that **HI 5** implies **HI 1**. This concludes the proof of the equivalence between the five properties.

Remarks. *X is hyperbolic if and only if X is hyperbolically imbedded in itself.*

This is obvious from **HI 1**.

If X_1 is hyperbolically imbedded in Y_1 and X_2 is hyperbolically imbedded in Y_2 then $X_1 \times X_2$ is hyperbolically imbedded in $Y_1 \times Y_2$.

The proof is immediate, using the fact that the projection on each factor of the product is Kobayashi distance decreasing.

Let $Y = \mathbf{P}^1(\mathbf{C})$ and $X = Y - \{\text{three points}\}$. Then X is hyperbolically imbedded in Y.

This follows from Theorem 3.3 of Chapter III, using the fact that the uiversal covering space of X is a disc; or using Borel's Theorem 1.1 of Chapter VII (cf. Theorem 2.2 of Chapter VII); or from Theorem 3.3 of Chapter IV, together with the construction of hyperbolic forms of Theorem 4.10 of Chapter IV, applied in the one-dimensional case.

Proposition 1.1. *Let X be a relatively compact complex subspace of a complex space Y. Then X is hyperbolically imbedded in Y if and only if, for every length function H on Y there is a constant $C > 0$ such that for all holomorphic maps $f: \mathbf{D} \to X$ we have*

$$f^*H \leqq CH_{\mathbf{D}}.$$

Proof. This is obvious, because over a compact subspace, all length functions are equivalent (each is less than a constant multiple of another).

Theorem 1.2. *Let X be a complex subspace, relatively compact in Y. Then X is hyperbolically imbedded in Y if and only if $\mathrm{Hol}(\mathbf{D}, X)$ is relatively locally compact in $\mathrm{Hol}(\mathbf{D}, Y)$.*

Proof. Assume $\mathrm{Hol}(\mathbf{D}, X)$ is relatively locally compact in $\mathrm{Hol}(\mathbf{D}, Y)$ but **HI 4** is false. Then for every length function on Y and every positive integer n there is a holomorphic map

$$f_n: \mathbf{D} \to X \quad \text{and} \quad z_n \in \mathbf{D}$$

such that $|df_n(z_n)v| \geqq n|v|$ for all $v \in T_{z_n}\mathbf{D}$. We may also assume $z_n = 0$. By relative compactness, there is some $y \in \bar{X}$ such that $f_n(0) \to y$. By assumption, after subsequencing, we can assume that $\{f_n\}$ converges uniformly on a neighborhood of 0. Then $f'_n(0)$ converges, which is a contradiction.

Conversely, assume X hyperbolically imbedded in Y. By Ascoli's theorem it suffices to prove that $\mathrm{Hol}(\mathbf{D}, X)$ is equicontinuous for a distance d_H arising from a length function on Y. But this is immediate from **HI 4**, thus concluding the proof of Theorem 1.2.

We say that X is **locally complete hyperbolic in** Y if for each point $y \in Y$ there exists a neighborhood U_y such that $U_y \cap X$ is complete hyperbolic.

Example. *Let D be a Cartier divisor on Y. Then $Y - D$ is locally complete hyperbolic in Y.*

Indeed, locally there is a polydisc $\mathbf{D}^n$ such that Y is a closed subspace of $\mathbf{D}^n$ and so is complete hyperbolic; and also D is the set of zeros of one analytic equation $f = 0$, so the local complement is complete hyperbolic by Proposition 3.7 of Chapter I.

Theorem 1.3. *Let X be a complex subspace of Y. Assume that*:

(a)   *X is relatively complete in Y;*
(b)   *X is hyperbolically imbedded in Y;*
(c)   *X is locally complete hyperbolic in Y.*

Then X is complete hyperbolic.

Proof. By **HI 4** we know that the inclusion $\pi: X \to Y$ is distance decreasing from d_X to d_H, so the theorem is a special case of Chapter I, Proposition 3.5.

Examples. I know of three basic examples of hyperbolic imbeddings. First in the 20's, Bloch and Cartan proved that the complement of certain hyperplanes in $\mathbf{P}^n$ is hyperbolically imbedded (in different terminology). Such theorems will be proved in Chapter VII and VIII.

Second, Kobayashi-Ochiai proved that the moduli space is hyperbolically imbedded in its minimal (Satake) compactification [K-O 1].

Third, Noguchi proved that for a proper family of curves of genus ≥ 2, a minimal model over the base is such that the complement of the bad fibers is hyperbolically imbedded in its closure [No 3]. The analogous question remains open when the generic fiber is a hyperbolic non-singular projective variety.

The second and third examples are beyond the techniques discussed in this book, and rely of course on much more algebraic geometry than we use here.

A generalization

Following Kiernan-Kobayashi [K-K 2], the notion of being hyperbolic and hyperbolically imbedded can be generalized in a way that is important for applications. Let Y be a complex space and let S be a closed complex subspace of Y. We define:

Y is **hyperbolic modulo** S if for every pair of distinct points x, y of Y not both contained S we have $d_Y(x, y) > 0$.

Y is **complete hyperbolic modulo** S if it is hyperbolic modulo S, and if for each sequence $\{y_n\}$ in Y which is d_Y-Cauchy we have one of the following:

CH$_S$ 1. $\{y_n\}$ converges to a point in Y.

CH$_S$ 2. For every open neighborhood U of S there exists some n_0 such that $y_n \in U$ for all $n \geq n_0$.

In the same vein, we deal with imbeddings.

Let X be a complex subspace of Y. We can stick an S in every one of the five conditions **HI 1** through **HI 5**, and denote these new conditions by **HI$_S$ 1** through **HI$_S$ 5**. We restate just two of these conditions here for simplicity, to give the idea.

HI$_S$ 3. Given a length function H on Y, there exists a continuous function $\varphi \geq 0$ on Y such that $\varphi > 0$ on $Y - S$ and

$$f^*(\varphi H) \leqq H_{\mathbf{D}}$$

for every holomorphic map $f: \mathbf{D} \to X$.

Note here the strict inequality in one instance, and the weak inequality in the other.

HI$_S$ 5. For every pair of distinct points $x, y \in \overline{X}$ not both contained in S there exist neighborhoods U of x and V of y in Y such that

$$d_X(U \cap X, V \cap X) > 0.$$

The proofs of equivalence are the same with or without the extra S, and define **hyperbolically imbedded modulo** S.

Similarly, let X be a complex subspace of Y. Let S be a closed complex subspace of Y. We say that $\mathrm{Hol}(\mathbf{D}, X)$ is **relatively locally compact in** $\mathrm{Hol}(\mathbf{D}, Y)$ **modulo** S if and only if given an infinite family $\mathscr{F}$ in $\mathrm{Hol}(\mathbf{D}, X)$ there is a subsequence $\{f_n\}$ such that: $\{f_n\}$ converges uniformly on compact subsets of $\mathbf{D}$ to an element of $\mathrm{Hol}(\mathbf{D}, Y)$; or given a neighborhood U of S in Y and a compact subset K of $\mathbf{D}$ there exists n_0 such that for all $n \geq n_0$ we have $f_n(K) \subset U$.

The following result corresponds to one direction in Theorem 1.2.

Theorem 1.4. *Let X be a relatively compact complex subspace of Y. Let S be a complex subspace of Y. If $\mathrm{Hol}(\mathbf{D}, X)$ is relatively locally compact in $\mathrm{Hol}(\mathbf{D}, Y)$ modulo S, then X is hyperbolically imbedded in Y modulo S.*

Proof. Assume that Hol($\mathbf{D}$, X) is relatively locally compact in Hol($\mathbf{D}$, Y) mod S. We prove that X is hyperbolically imbedded in Y mod S following [K-K 2]. Let x, $y \in \bar{X}$ and say $x \notin S$. Then there exist open neighborhoods U of x and V of y in Y such that

$$\bar{U} \cap \bar{V} \quad \text{is empty} \quad \text{and} \quad \bar{U} \cap S \quad \text{is empty.}$$

Taking U small enough we may assume that U is hyperbolic. Let W be a still smaller neighborhood of x such that $\bar{W}$ is compact and $\bar{W} \subset U$. We claim that:

There exists $r < 1$ such that if $f: \mathbf{D} \to X$ is holomorphic and $f(0) \in \bar{W}$ then $f(\mathbf{D}_r) \subset U$.

For otherwise, given a positive integer n there exists $f_n: \mathbf{D} \to X$ holomorphic such that $f_n(0) \in \bar{W}$ but $f_n(\mathbf{D}_{1/n}) \not\subset U$. Then the sequence $\{f_n\}$ has no subsequence satisfying either possibility in the definition of Hol($\mathbf{D}$, X) being relatively locally compact in Hol($\mathbf{D}$, Y) mod S, which is a contradiction.

Let U_1 be an open neighborhood of x in Y such that

$$x \in U_1 \subset \bar{U}_1 \subset W \subset \bar{W}.$$

We shall prove that $d_X(U_1 \cap X, V \cap X) > 0$. Let

$$b = d_U(\bar{U}_1, U - W) > 0$$

because U is hyperbolic and d_U induces the ordinary topology. Let $c > 0$ be such that

$$d_{\mathbf{D}}(0, z) \geqq c \cdot d_{\mathbf{D}_r}(0, z) \qquad \text{for all} \quad z \in \mathbf{D}_{r/2}.$$

Let $x_1 \in U_1 \cap X$ and $y_1 \in V \cap X$. It will now suffice to prove that $d_X(x_1, y_1) \geqq bc$. Let $f_i: \mathbf{D} \to X$ ($i = 1, \ldots, m$) be a Kobayashi chain of discs joining x_1 to y_1 in X. Without loss of generality, we may suppose that the corresponding sequence of pairs of points (p_i, q_i) in $\mathbf{D}$ are $(0, z_i)$, and after inserting additional points in the chain, we may also suppose that $z_i \in \mathbf{D}_{r/2}$ and that for some $k > 1$ we have

$$f_i(0) \in W \qquad \text{for} \quad i = 1, \ldots, k,$$
$$f_k(z_k) \notin W.$$

Then

$$\sum_{i=1}^{m} d_{\mathbf{D}}(0, z_i) \geqq \sum_{i=1}^{k} d_{\mathbf{D}_r}(0, z_i) \geqq c \sum_{i=1}^{k} d_{\mathbf{D}_r}(0, z_i)$$

$$\geqq c \sum_{i=1}^{k} d_U(f_i(0), f_i(z_i))$$

$$\geqq c \cdot d_U(x_1, f_k(z_k)) \geqq cb.$$

This proves $d_X(x_1, y_1) \geqq cb$ and concludes the proof of the desired implication. The argument was essentially of the same type as Lemma 2.5 of Chapter I.

I would conjecture the converse, but as in Problem 2 of Kiernan–Kobayashi [K–K 2], p. 215 I did not see how to prove it. The notion of Hol($\mathbf{D}, X$) being relatively locally compact mod S is a variation on Wu's "tautness" used in [K–K 2]. It stems from the example of Cartan's theorem in Chapter VIII.

The proof of Theorem 1.3 also applies essentially without change to yield:

Theorem 1.5. *Let X be a complex subspace of Y and let S be a closed complex subspace. Assume:*

(a)   *X is relatively complete in Y.*
(b)   *X is hyperbolically imbedded modulo S in Y.*
(c)   *X is locally complete hyperbolic in Y.*

Then X is complete hyperbolic modulo S.

The exceptional set S does occur in applications. See for instance Chapter III, §3, where we recall theorems of Bloch, Cartan, and Green, giving concrete examples of the above notions; as well as Chapters VII and VIII, where we prove them.

II, §2. KWACK'S THEOREM (BIG PICARD) ON D

The Big Picard theorem states that a holomorphic function on the punctured disc having an essential singularity at 0 misses at most one value in **C**. We view such a function as a map into the projective line $\mathbf{P}^1$ minus two points. Then we can state the theorem in the form: a holomorphic mapping

$$\mathbf{D}^* \to \mathbf{P}^1 - \{\text{three points}\}$$

has a removable singularity at 0, that is extends holomorphically to **D**. This theorem extends to higher dimensional contexts for the image of f, or even the domain of f. In this section, we still look at the domain **D***, but generalize the image suitably. This section reproduces Kwack's extension of the Big Picard theorem [Kw].

Throughout this section, we let X be a complex subspace of Y, and let

$$f : \mathbf{D}^* \to X$$

be a holomorphic map of the punctured disc into X.

We are interested when this map extends to a holomorphic map on all of **D**, that is extends to the origin. In other words, we seek conditions when f has a removable singularity at 0. Consider the following conditions:

KW 1. X is hyperbolically imbedded in Y, and there exists a sequence $\{z_k\}$ in **D*** such that $z_k \to 0$ and $\{f(z_k)\}$ converges to a point $y \in \bar{X}$.

Note: The condition about the existence of the sequence $\{z_k\}$ is automatically satisfied if X is relatively compact in Y.

KW 2. X is hyperbolically imbedded in Y, and there exists a sequence of positive numbers $\{r_k\}$ decreasing to 0, such that if $\mathbf{S}(r_k) = \mathbf{S}(0, r_k)$ is the circle of radius r_k, then $f(\mathbf{S}(r_k))$ converges to some $y \in \bar{X}$.

KW 3. The map f extends to a holomorphic map in a neighborhood of 0, i.e. f has a removable singularity at 0.

Theorem 2.1. *We have the implications*

$$\mathbf{KW\ 1} \quad \Rightarrow \quad \mathbf{KW\ 2} \quad \Rightarrow \quad \mathbf{KW\ 3}.$$

In particular, if X is hyperbolically imbedded in Y, and $\bar{X}$ is compact, then every holomorphic map

$$f : \mathbf{D}^* \to X$$

extends to a holomorphic map of $\mathbf{D}$ into Y.

Proof. Assume **KW 1**. Let $r_k = |z_k|$. Let U be a hyperbolic neighborhood of y such that its closure $\bar{U}$ is compact. Such a neighborhood can be found, for instance because locally near y, Y is a closed subspace of a polydisc in some $\mathbf{C}^N$. To show **KW 2**, it will suffice to prove that

$$f(\mathbf{S}(r_k)) \subset U \quad \text{for } k \text{ sufficiently large.}$$

Suppose this is not the case. Then for arbitrarily large k there is a point $z'_k \in \mathbf{S}(r_k)$ such that $f(z'_k) \notin U$. From the continuity of a distance on Y defining its topology, we can find z'_k such that $f(z'_k)$ lies in a fixed compact set, or even in $\bar{U} - U$, for instance. In particular, passing to a subsequence, we may assume that $f(z'_k)$ converges to a point $y' \in \bar{X}$. We then have $y' \neq y$ since $f(z'_k) \notin U$. We also have the inequality

$$d_X(f(z_k), f(z'_k)) \leq d_{\mathbf{D}^*}(z_k, z'_k) \to 0 \qquad \text{as} \quad k \to \infty,$$

because the circle of radius r has hyperbolic length $O(-1/\log r)$. One sees this by representing $\mathbf{D}^*$ as quotient of the upper half plane, where the segment $-\frac{1}{2} \leq x \leq \frac{1}{2}$, y constant, maps on the circle of radius $e^{-2\pi y}$, and $ds^2 = (dx^2 + dy^2)/y^2 = dx^2/y^2$ on this segment. This contradicts the hypothesis that X is hyperbolically imbedded in Y, and concludes the proof that **KW 1** implies **KW 2**.

Next we prove that **KW 2** implies **KW 3**. We let U be a neighborhood of y, identified with a subspace of a polydisc in $\mathbf{C}^N$, such that the closure $\bar{U}$ of U in Y is compact and contained in the polydisc.

It will suffice to prove that there exists $c > 0$ such that $f(\mathbf{D}_c^*) \subset U$, thus showing that the coordinate functions of f are bounded near 0, so 0 is a removable singularity. Assume there is no such c. After renumbering the sequence, we can assume that $f(\mathbf{S}(r_k)) \subset U$ for all k. Let a_k, b_k be positive numbers with

$$a_k < r_k < b_k$$

such that the annulus A_k defined by $a_k < |z| < b_k$ is the largest annulus whose image under f is contained in U. We let

$$\alpha_k(t) = a_k e^{2\pi i t} \quad \text{and} \quad \beta_k(t) = b_k e^{2\pi i t}, \qquad 0 \leq t \leq 1$$

be the two circles bounding the open annulus. Then

$$f(\alpha_k) \quad \text{and} \quad f(\beta_k) \subset \bar{U},$$

but these images of the two circles α_k and β_k are not contained in U. Since the hyperbolic lengths of the circles of radius a_k and b_k tend to 0 as $k \to \infty$, and since f is distance decreasing from $d_{\mathbf{D}^*}$ to d_X, it follows that the d_X-diameters of $f(\alpha_k)$ and $f(\beta_k)$ tend to 0. By **HI 4**, so tend the d_H-diameters. Since $\bar{U}$ is compact, after subsequencing we may assume that $f(\alpha_k)$ and $f(\beta_k)$ converge to points y' and y'' respectively, in $\bar{U} - U$. Then we have $y' \neq y$ and $y'' \neq y$. We let z_k be a point on $\mathbf{S}(r_k)$ so $z_k \to y$ as $k \to \infty$.

Let $(f_1,\ldots,f_N)$ be the coordinate functions of f as a map of U into $\mathbf{C}^N$. Without loss of generality, we may assume that

$$\lim_{k\to\infty} f_1(\alpha_k) = y_1' \neq 0,$$

$$\lim_{k\to\infty} f_1(\beta_k) = y_1'' \neq 0,$$

$$\lim_{k\to\infty} f_1(z_k) = y_1 = 0.$$

In other words, $y = (y_1,\ldots,y_N)$ is the origin in the polydisc, and the first coordinate of f distinguishes y' and y'' from y. It follows that for all $k \geq k_0$ we have

$$f_1(z_k) \notin f_1(\alpha_k) \cup f_1(\beta_k).$$

In other words, $f_1(z_k)$ is not in the image of the two circles α_k, β_k under f. Thus we can find a simply connected neighborhood of $f_1(\alpha_k) \cup f_1(\beta_k)$ which does not intersect a small disc in $\mathbf{C}$ centered at 0.

We now compute "winding numbers". Let $(w_1,\ldots,w_N)$ be the coordinate functions on $\mathbf{C}^N$ so that $f_1 = w_1 \circ f$. In light of the above neighborhoods, for large k we find:

$$\int_{f(\alpha_k)} d\log(w_1 - f_1(z_k)) = 0 = \int_{\alpha_k} d\log(w_1 \circ f - f_1(z_k))$$

and similarly with β_k instead of α_k. On the other hand,

$$\frac{1}{2\pi\sqrt{-1}}\int_{\alpha_k - \beta_k} d\log(w_1 \circ f - f_1(z_k))$$

   = number of zeros of $w_1 \circ f$ — number of poles of $w_1 \circ f$ inside
      the annulus A_k
   ≥ 1

because the functions have no poles, and at least one zero at z_k. This contradiction finishes the proof of Kwack's theorem.

Remark. The argument with the winding numbers stems from Grauert-Reckziegel [G–R], who proved a similar extension theorem for sections. As usual, these authors work with a definition of hyperbolicity in terms of "negative curvature", but formally, the situation is the same as for Kobayashi hyperbolicity. The context of sections is not needed to formulate or prove the extension theorem, which is of "Big Picard" type. In §4 we give a formulation of a result of Noguchi [No 3] in the context of Kobayashi hyperbolicity, and without the context of sections. The formulation for sections will be given in Chapter III, Theorem 3.2. The emphasis on sections stems from the motivation of Mordell's conjecture in the function field case, i.e. for sections of analytic or algebraic families, cf. [La 0], [Ma], [Gr 1], and [G-R].

Actually Kwack dealt with the case when $X = Y$ is compact. It was Kobayashi who introduced the notion of being hyperbolically imbedded [Ko 3]. We have given the more general version right away.

After handling one-dimensional applications of Kwack's theorem in §4, we shall give the K^3-higher dimensional version in §5, which sets the context for the higher dimensional version of Noguchi's theorem.

II, §3. SOME RESULTS IN MEASURE THEORY

We assume known some basic facts of measure theory. In particular, if Ω is a **volume form**, that is a positive differential form of top degree on an oriented real manifold, then Ω induces a positive functional on the space of continuous functions with compact support (integrating against Ω), and thereby a unique positive measure on the Borel sets. Cf. the last part of my *Real Analysis* for instance to see the details. I do not identify the form of top degree with the corresponding measure.

In this section we prove some results to be used in the next sections in the context of hyperbolicity. These results have to do with measure theory in various contexts. They involve both dimension 1 and higher dimensions. We shall first mention the results in dimension 1 and then extend them.

1-dimensional complex results

We let $\mathbf{D}^*$ be the punctured disc $\mathbf{D} - \{0\}$. Under the map

$$z \mapsto e^{2\pi i z}$$

the upper half plane covers $\mathbf{D}^*$. By the **hyperbolic measure** in the upper half plane, we mean the measure given by

$$\frac{dx\,dy}{y^2} \quad \text{if} \quad z = x + iy.$$

By the **hyperbolic measure** $\mu_{\mathbf{D}^*}$ on $\mathbf{D}^*$ we mean the measure induced from the measure on the upper half plane. We shall explain later in this section how this measure is associated with the hyperbolic metric. For a start, we just want to give one useful result.

Proposition 3.1. *Let $\mu_{\mathbf{D}*}$ be the hyperbolic measure on $\mathbf{D}^*$. Then*

$$\mu_{\mathbf{D}*}(\mathbf{D}^*_{1/k}) \to 0 \qquad as \quad k \to \infty.$$

Proof. The upper part of a strip $-\frac{1}{2} < x \leqq \frac{1}{2}$, $y > C$ is mapped bijectively on a punctured disc of radius $e^{-2\pi C}$, and the measure of the top part of the strip goes to 0 as C tends to infinity. This proves the proposition.

The next result has to do with the area of a 1-dimensional complex subspace of $\mathbf{C}^N$. A much more general result will be proved afterward, but the 1-dimensional case is valuable for its own sake, and I want to illustrate a different method. The reader who is interested only in an efficient general treatment can omit the rest of this 1-dimensional discussion.

Let ω be the hermitian positive $(1, 1)$-form on $\mathbf{C}^N$, so

$$\omega(z) = \sum_{j=1}^{N} \frac{\sqrt{-1}}{2} \, dz_j \wedge d\bar{z}_j.$$

If A is a complex subspace of dimension 1, then the restriction of ω to A_{reg} (regular points) is a volume form on A_{reg}, and thus defines a measure on A_{reg}. The singularities of A can consist only of isolated points, which have measure 0. This is the simple case which arises in the next theorem. We let μ_{euc} denote the measure induced on A by the above form.

We shall map a disc into a ball, such that some point of the disc goes to the center of the ball, and the boundary of the disc goes completely outside the ball. Then we shall see that the 2-dimensional area of the image of the disc contained in the ball is at least equal to what it would be if the map is linear. Note that this type of statement is totally false in the C^∞ category.

As usual, we let $\mathbf{D}_a$ be the disc of radius a and $\mathbf{S}_a$ the circle of radius a.

Theorem 3.2. *Let $c > 0$. Let $x_0 \in \mathbf{C}^N$. Let U be an open set in $\mathbf{C}$ containing the disc $\bar{\mathbf{D}}_a$ for some $a > 0$. Let*

$$f: U \to \mathbf{C}^N, \qquad f = (f_1, \ldots, f_N)$$

be holomorphic. Suppose that there is a point $z_0 \in \mathbf{D}_a$ such that $f(z_0) = x_0$. On the other hand, suppose that $f(\mathbf{S}_a)$ lies outside the ball $\bar{\mathbf{B}}(x_0, c)$. Then the euclidean measure of

$$f(\mathbf{D}_a) \cap \mathbf{B}(x_0, c)$$

satisfies

$$\mu_{\text{euc}}(f(\mathbf{D}_a) \cap B(x_0, c)) \geq \pi c^2.$$

Proof. I owe the following proof when f is injective to Peter Jones. After making a dilation if necessary, we may assume

$$a = c = 1.$$

We can also assume $x_0 = 0$ in $\mathbf{C}^N$ and $f(0) = 0$.

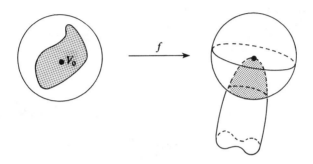

We let $V = f^{-1}\mathbf{B}$ where $\mathbf{B}$ is the unit ball, and we let V_0 be the connected component of V in $\mathbf{D}$ containing the origin.

We claim that V_0 is simply connected. Let γ be a closed curve in V_0. Then γ is homotopic to a piecewise linear path close by, say with vertical and horizontal sides; and we can decompose γ into a sum of simple closed curves. It suffices to show that each such curve η is deformable to a point in V_0. We have $\sum |f_j(z)|^2 \leq 1$ on the boundary of η. Since $\sum |f_j(z)|^2$ is subharmonic (see below), and therefore satisfies the maximum principle, it follows that this inequality is valid for z in the interior of η, which is therefore contained in V_0. This proves the simple connectedness.

By the Riemann mapping theorem, there exists a holomorphic isomorphism of V_0 with $\mathbf{D}$. Composing this mapping with f, we are reduced to proving the following normalized version of Theorem 3.2.

Lemma 3.3. *Let*

$$f = (f_1, \ldots, f_N) \colon \bar{\mathbf{D}} \to \mathbf{C}^N$$

be holomorphic and injective. Assume that f maps the boundary of $\mathbf{D}$ outside the unit ball, i.e. $|f(z)| \geq 1$ for $|z| = 1$, and $f(0) = 0$. Then

$$\pi \leq \mu_{\text{euc}}(f(\mathbf{D})).$$

Proof. Let

$$f_j(z) = \sum_{k=1}^{\infty} a_{j,k} z^k.$$

Let ω be the euclidean $(1, 1)$-form of $\mathbf{C}^N$. Then

$$\mu_{\mathrm{euc}}(f(\mathbf{D})) = \int_{\mathbf{D}} f^*(\omega) = \int_{\mathbf{D}} \sum_{j=1}^{N} |f'_j(z)|^2 \, d\mu(z)$$

(where μ is the Lebesgue measure on the disc)

$$= \pi \sum_{j=1}^{N} \sum_{k=1}^{\infty} k |a_{j,k}|^2$$

$$\geqq \pi \sum_{j=1}^{N} \sum_{k=1}^{\infty} |a_{j,k}|^2$$

$$= \pi \int_0^{2\pi} \sum_{j=1}^{N} |f_j(e^{i\theta})|^2 \frac{d\theta}{2\pi}$$

$$\geqq \pi, \quad \text{because} \quad |f(z)|^2 \geqq 1.$$

This proves the desired inequality.

For the convenience of the reader, I reproduce the details of what we have used about subharmonic functions, which is practically nothing. For our purposes, we say that a real-valued function is **subharmonic** if it satisfies the inequality

$$\varphi(a) \leqq \int_0^{2\pi} \varphi(a + re^{i\theta}) \frac{d\theta}{2\pi}$$

for any radius r such that the closed disc is contained in the open set of definition. The subharmonic functions obviously form a positive cone. If f is holomorphic, then by Cauchy's theorem,

$$f(a) = \int_0^{2\pi} f(a + re^{i\theta}) \frac{d\theta}{2\pi},$$

and we can also apply this to f^2 instead of f. Taking absolute values we see that $\varphi = f\bar{f} = |f|^2$ is subharmonic.

A subharmonic function satisfies the maximum principle. In other words, if φ is subharmonic on an open set U, and $a \in U$, then for all r suffi-

ciently small there is a point b on the circle of radius r such that
$\varphi(b) \geqq \varphi(a)$.

For otherwise, there is some r such that $\varphi(a) > \varphi(b)$ for all points b on the circle of radius r, and this contradicts the previous integral inequality.

As usual, we get:

If φ is subharmonic on a relatively compact open set U, and extends continuously on the boundary, then φ has a maximum on the boundary of U.

Higher dimensional complex results

We are now through with the 1-dimensional statements. Next we shall recall certain extensions to higher dimension. Although these will also be applied essentially in dimension 1, nevertheless the statements we give hold in greater generality and are necessary for the application. For instance, the proof of Theorem 3.2 held only if f is injective. Other arguments provide a proof in general, as we shall see. Working in higher dimensions in some sense forces the proofs to be "natural".

Let M be a smooth submanifold of $\mathbf{R}^n$, and $\dim_{\mathbf{R}} M = k$.

First recall that the euclidean scalar product on $\mathbf{R}^n$ gives rise naturally to a scalar product on the exterior powers, and if E is a real linear subspace of $\mathbf{R}^n$ we get an induced scalar product on E and its exterior powers. Thus we get a natural volume form on E. This is applied when

$$E = T_{x,M} \subset \mathbf{R}^n,$$

so we get the volume form $\Omega_{k,x}$ on $T_{x,M}$ for each $x \in M$, whence the volume form Ω_k on M. We let μ_k be the associated measure, cf. my *Real Analysis*, Chapter XX, or [La 4], Ch. IX, §3.

Proposition 3.4. *Let $k \leqq n$ be positive integers. Let M be a C^∞ submanifold of a neighborhood of 0 in $\mathbf{R}^n$, of dimension k, such that $0 \in M$. Let $| \ |$ be the euclidean norm and let*

$$M_r = \{x \in M \text{ such that } |x| < r\}.$$

Let v_k be the Lebesgue volume of the unit ball in $\mathbf{R}^k$. Then

$$\lim_{r \to 0} \frac{1}{v_k r^k} \mu_k(M_r) = 1.$$

Proof. Let $x = (x_1, \ldots, x_k)$. We can choose coordinates in $\mathbf{R}^n$ such that M locally near 0 is parametrized as

$$x \mapsto (x, h(x)), \qquad \text{so } x_j = h_j(x) \text{ for } j = k+1, \ldots, n,$$

and such that $T_{0,M}$ can be identified with $\mathbf{R}^k \times 0^{n-k} \subset \mathbf{R}^n$. Thus we have

$$h(0) = 0 \qquad \text{and} \qquad dh(0) = 0.$$

It follows that

$$h(x) = O(|x|^2) \qquad \text{as} \quad |x| \to 0.$$

Write $\varphi(x) = (x, h(x))$. Then

$$d\varphi(x) = I + A(x),$$

where $I: \mathbf{R}^k \to \mathbf{R}^k \times 0^{n-k}$ is the natural imbedding of $\mathbf{R}^k$ in $\mathbf{R}^n$, and

$$A(x) = O(|x|) \qquad \text{as} \quad |x| \to 0,$$

say using the sup norm on the coefficients of $A(x)$ viewed as a matrix, or taking the norm as a linear map $\mathbf{R}^k \to \mathbf{R}^n$, whatever you want. Then the pull-back of Ω_k to $\mathbf{R}^k$ near the origin is of the form

$$(\textstyle\bigwedge^k d\varphi(x))^*(\Omega_{k,x}) = (1 + O(|x|^2))\, dx_1 \wedge \cdots \wedge dx_k.$$

Furthermore, the euclidean distances are also nearly equal, that is

$$|x|_k = |(x, h(x))|_n + O(|x|^2) \qquad \text{as} \quad |x| \to 0.$$

Hence asymptotically, the measure of the $\mathbf{B}_r^k$ in $\mathbf{R}^k$ is the same as the measure of M_r with respect to μ_k. This concludes the proof.

We return to a complex situation. Let

$$\omega(z) = \frac{\sqrt{-1}}{2} \sum h_{ij}\, dz_i \wedge d\bar{z}_j$$

be a positive $(1, 1)$-form on an open set U of $\mathbf{C}^N$, where (h_{ij}) is a hermitian positive definite matrix. This form defines a Riemannian metric and higher degree forms as follows. The associated hermitian metric is given in terms of the complex coordinates by

$$h_\omega = \sum h_{ij}\, dz_i\, d\bar{z}_j.$$

Applying the Gram–Schmidt orthogonalization process with respect to this hermitian metric, we can write

$$h_\omega = \sum \varphi_i \otimes \bar{\varphi}_i,$$

where $\{\varphi_i\}$ is an orthonormal basis for the form. Each φ_i is a linear combination of $dz_1, \ldots, dz_N$ with C^∞ coefficients. Write

$$\varphi_i = \alpha_i + \sqrt{-1}\,\beta_i,$$

where α_i and β_i are real forms. By definition, the **associated Riemannian metric** is defined to be

$$\mathrm{Re}(h_\omega) = \sum_{i=1}^{N} \alpha_i^2 + \beta_i^2 = ds^2,$$

which can be shown to be independent of the choice of orthonormal basis.

Again by definition, the **volume form** in $\mathbf{C}^N$ for this Riemannian metric is the differential form

$$\alpha_1 \wedge \beta_1 \wedge \cdots \wedge \alpha_N \wedge \beta_N.$$

Then we have the associated measure μ.

Next let X be a complex subspace of $\mathbf{C}^N$, of pure dimension m. We suppose that $\bar{X}$ is compact, and $\bar{X} \subset U$. We then have two possible volume forms on the complex submanifold $M = X_{\mathrm{reg}}$ of $\mathbf{C}^N$.

(a) The $(1, 1)$-form ω induces a $(1, 1)$-form ω_M on M, whence a hermitian metric on M, whence a Riemannian metric, whence a volume form and a measure, which we may denote by

$$\mu_{\omega, M} \quad \text{or} \quad \mu_{\omega, 2m}.$$

In the second notation, we leave out the M but specify the real dimension. This measure is uniquely determined by the original ω. If ω is the euclidean form

$$\omega = \frac{\sqrt{-1}}{2} \sum dz_i \wedge d\bar{z}_j,$$

then we simply write μ_{2m}, and call it the **euclidean measure** in real dimension $2m$. There is one such measure for each complex submanifold of dimension m in $\mathbf{C}^N$.

(b) The form

$$\omega_m = \Omega = \omega^m/m!$$

is a positive (m, m)-form, which induces a volume form on X_{reg}. Again resolution of singularities shows that locally, X_{reg} has finite volume with respect to this form. In other words, if $\bar{\mathbf{B}}_R$ is a closed ball contained in U, then

$$\mu_\Omega(X_{\mathrm{reg}} \cap \mathbf{B}_R) \quad \text{is finite},$$

where μ_Ω is the measure associated with Ω, restricted to X_{reg}, cf. again *Real Analysis*, Chapter XX. [Those who don't like to use resolution and want to see the technique of projecting on affine space can refer to Stolzenberg [Stolz] or Griffiths-Harris, the Proposition, p. 32.]

Theorem 3.5 (Wirtinger's theorem). *Let M be a complex submanifold of $\mathbf{C}^N$. Let ω be a positive $(1, 1)$-form as above. Let $m = \dim M$ and let μ be the measure on M induced by the Riemannian metric corresponding to ω. Then*

$$\mu(M) = \int_M \omega^m/m!.$$

Proof. The assertion is local, so we may assume that M is open in $\mathbf{C}^N$ and the notation is that preceding the theorem. Then

$$\omega = \frac{\sqrt{-1}}{2} \sum_{i=1}^N \varphi_i \wedge \bar{\varphi}_i,$$

so

$$\omega^N = \left(\frac{\sqrt{-1}}{2}\right)^N N! \, \varphi_1 \wedge \bar{\varphi}_1 \wedge \cdots \wedge \varphi_N \wedge \bar{\varphi}_N$$

$$= N! \, \alpha_1 \wedge \beta_1 \wedge \cdots \wedge \alpha_N \wedge \beta_N$$

as was to be shown.

Suppose that X is a complex subspace of $\mathbf{C}^N$. For $r > 0$ we let

$$X_r = X \cap \mathbf{B}_r,$$

where $\mathbf{B}_r = \mathbf{B}(0, r)$ is the ball of radius r. Let $\mathbf{S}_r = \mathbf{S}(0, r)$ be the sphere. Consider the function

$$q(x) = |x|^2 \qquad \text{for} \quad x \in X_{\mathrm{reg}}.$$

Then q is real a analytic. Except for isolated values of r, the level space

$$X_{\text{reg}} \cap S_r$$

is a real analytic manifold, of real dimension $2m - 1$. Note that the singularities of X have real dimension $\leq 2m - 2$. So even if we intersect $X \cap S_r$ rather than $X_{\text{reg}} \cap S_r$ for regular values of r, the difference is still of lower dimension.

An integral over a complex space will always mean over its regular points.

In the proof of the next theorem, we shall use Stokes' theorem applied to the regular part. For the justification that the usual formalism works, because the singularities are "negligible", cf. my *Real Analysis*, Chapter 20, §6, or [La 4], Chapter IX, §3.

Theorem 3.6 (Lelong [Le 1] and [Le 2]). *Let μ_{2m} be the euclidean measure induced on an m-dimensional complex subspace X of $\mathbf{C}^N$. Let $\bar{X}$ be the closure, and assume that $\bar{X} - X$ lies outside the ball $\mathbf{B}(R_0)$. Assume that $0 \in X$. Then for $r < R_0$ we have*

$$\mu_{2m}(X_r) \geqq v_{2m} r^{2m},$$

where v_{2m} is the Lebesgue volume of the unit ball in $\mathbf{R}^{2m}$. In fact, if we put

$$F(r) = \mu_{2m}(X_r)/r^{2m},$$

then F is an increasing function of r.

Proof. We shall first prove the second statement, that F is increasing. We use the differential forms:

$$\varphi = dd^c |z|^2 = \frac{\sqrt{-1}}{2\pi} \sum dz_i \wedge d\bar{z}_i = \frac{\sqrt{-1}}{2\pi} \langle dz, dz \rangle,$$

$$\omega = dd^c \log|z|^2 = \frac{\sqrt{-1}}{2\pi} \frac{\langle z, z \rangle \langle dz, dz \rangle - \langle dz, z \rangle \langle z, dz \rangle}{|z|^4},$$

where $\langle \ , \ \rangle$ is the standard scalar hermitian product, and

$$d^c = \frac{1}{4\pi\sqrt{-1}} (\partial - \bar{\partial}).$$

The notation $\langle dz, z \rangle$ means $\sum \bar{z}_i \, dz_i$ and similarly for $\langle z, dz \rangle$. The first form φ is the euclidean form, up to a normalizing constant factor; and the second form ω is called the **Fubini–Study form**. It is the pull-back to $\mathbf{C}^N - \{0\}$ of a positive $(1, 1)$-form on $\mathbf{P}^{N-1}$. But never mind this here. I am looking for ad hoc efficiency.

We have $d\varphi = 0$ and so $d\varphi^k = 0$ for positive integers k. Hence

$$\varphi^m = d(d^c |z|^2 \wedge \varphi^{m-1}).$$

Similarly for ω. We observe that on the tangent space to the sphere $|z| = r$, we have immediately the identities

$$\frac{1}{r^2} d^c |z|^2 = d^c \log |z|^2,$$

$$\frac{1}{r^2} dd^c |z|^2 = dd^c \log |z|^2, \qquad \text{because} \quad d|z|^2 = 0.$$

By Wirtinger's theorem we have

$$\mu_{2m}(X_r) = C_m \int_{X_r} \varphi^m,$$

with an appropriate constant C_m. By Stokes' theorem and the above identities, we find:

$$
\begin{aligned}
\frac{1}{C_m}[F(R) - F(r)] &= \frac{1}{R^{2m}} \int_{X_R} \varphi^m - \frac{1}{r^{2m}} \int_{X_r} \varphi^m \\
&= \frac{1}{R^{2m}} \int_{\partial X_R} d^c |z|^2 \wedge \varphi^{m-1} - \frac{1}{r^{2m}} \int_{\partial X_r} d^c |z|^2 \wedge \varphi^{m-1} \\
&= \int_{\partial X_R} d^c \log |z|^2 \wedge \omega^{m-1} - \int_{\partial X_r} d^c \log |z|^2 \wedge \omega^{m-1} \\
&= \int_{X(r, R)} \omega^m,
\end{aligned}
$$

where $X(r, R)$ is the intersection of X with the annulus $r < |z| < R$.

Since ω is a positive $(1, 1)$-form, it follows that ω^m is a volume form on X_{reg}, whence the right-hand side is ≥ 0. This concludes the proof that F is increasing.

If $0 \in X_{\text{reg}}$ then the first statement follows from this and Proposition 3.4. Suppose 0 is not a regular point of X. Then we can find a sequence of regular points $\{x_k\}$ converging to 0. Applying the above result to these cases concludes the proof of the theorem.

Remarks. Lelong ([Le 2], p. 70) attributes the origin of the two forms φ and ω to Poincaré, and to De Rham–Kodaira for a "modern formulation". The theorem is basic in many ways, and has been used many times, cf. Griffiths [Gri 2], p. 13, [Stoll 1], and Stolzenberg [Stolz], for instance. The proof in Stolzenberg (attributed to Federer) is less elegant than Lelong's original proof.

Aside from the history, there is a mathematical point involved in the last argument when the origin is not a regular point. In that case, Proposition 3.4 can be generalized, and the asymptotic value is the multiplicity of the complex space at the origin. This is also due to Lelong [Le 1], cf. [Gr 2], following the same sort of argument.

Hausdorff measures

Next we discuss Hausdorff measures.

Let X be a metric space with distance function d. Let p be a real number with $0 \leq p < \infty$. Let $c > 0$. Let A be a subset of X. We define

$$\mu_{d,c}^p(A) = \inf \sum_{i=1}^{\infty} v_p \, \text{diam}(A_i)^p / 2^p,$$

where the inf is taken over all sequences $\{A_i\}$ of sets of diameter $\text{diam}(A_i) \leq c$ covering A. We have normalized the expression on the right by the factor $v_p/2^p$ where v_p is the Lebesgue volume of the unit ball in $\mathbf{R}^p$, and the 2^p is thrown in because the radius is half the diameter. Also we should write $\mu_{d,c,X}^p$ with X in the notation.

Instead of A_i we could also take balls B_i, but it is technically convenient to allow for more general sets. See [Sm].

It is easily proved that $\mu_{d,c}^p$ is an outer measure on the subsets of X. It is called the **Hausdorff outer measure**.

Let Y be a subset of X, with the induced distance. Then the Hausdorff outer measure of subsets of Y is the same whether taken with respect to Y or with respect to X.

This is obvious, and shows that specifying X in the notation is not really necessary. This statement would however not be true if we used balls instead of arbitrary sets in the definition.

The outer measure $\mu_{d,c}^p$ has almost no measurable sets. But we define the p-**Hausdorff measure** μ_d^p by

$$\mu_d^p(A) = \lim_{c \to 0} \mu_{d,c}^p(A).$$

The following theorem is one of the basic results of measure theory.

Theorem 3.7. *Open and closed sets are (Hahn) measurable for the Hausdorff measure, which is therefore a positive measure on the Borel sets, and is a regular measure.*

For a sketch of proof, see my *Real Analysis*, Chapter XI, Exercises 21, 22, 23. But generally, for the basic theorems on Hausdorff measures, see Smith [Sm], Chapter XIII, §3, §4 and Chapter XV, Theorem 6.2.

Once we know about the measurability of Borel sets, the following assertion is obvious:

Let $f: X \to Y$ be a continuous map, distance decreasing with respect to distances d'_X and d'_Y. Suppose also that f maps Borel sets to Borel sets. Then f is measure decreasing for the associated p-Hausdorff measures.

This assertion will be applied to holomorphic maps from one complex space to another.

Theorem 3.8. *Let μ_{euc} be the Lebesgue measure on $\mathbf{R}^n$. If d is the euclidean distance, then on Borel sets of $\mathbf{R}^n$,*

$$\mu_{\text{euc}} = \mu_d^n.$$

The exact equality is a little elaborate to prove. In the applications, we shall only need that each side is less than a constant times the other, which is an easy exercise. One uses that

$$\mu_{\text{euc}}(A) = \inf_U \mu_{\text{euc}}(U),$$

where the inf is taken over open sets containing A. Also one observes that the equality to be proved is a local relation.

Theorem 3.9. *Let M be a hermitian manifold, with its positive $(1, 1)$-form ω, and volume form $\Omega = \omega^m/m!$, where $m = \dim M$. Let μ_Ω be the measure associated with Ω, that is the unique regular measure such that on open subsets V of M we have*

$$\mu_\Omega(V) = \int_V \Omega.$$

Let d be the Riemannian distance function, and μ_d^{2m} the Hausdorff measure. Then

$$\mu_\Omega = \mu_d^{2m}.$$

Proof. One can follow all the steps of the proof of Theorem 3.8, and carry them out in the Riemannian case to get the desired proof. But there is a neat way to go around that if one hits the problem with Nash's Imbedding Theorem, which says that a Riemannian manifold has a metric imbedding in some euclidean space of suitably high dimension $\mathbf{R}^N$. If x, $y \in M$ and d_M, d_{euc} denote the distances in M and in $\mathbf{R}^N$ with respect to the euclidean length function on $\mathbf{R}^N$, and its restriction to M, then

$$\lim_{y \to x} d_{euc}(x, y)/d_M(x, y) = 1.$$

One can then quote Theorem 3.8 to conclude the proof, because the $2m$-th Hausdorff measure on M is the restriction of the $2m$-th Hausdorff measure on $\mathbf{R}^N$. Of course, the theorem can be similarly stated for a Riemannian manifold.

Example. On the upper half plane, which is isomorphic to the disc, we have the Kobayashi metric corresponding to the Kobayashi metric on the disc **D**, and called the hyperbolic metric. The punctured disc admits the upper half plane as universal covering space. As we saw at the beginning of the section, it then inherits this hyperbolic metric, and its corresponding distance function $d_{\mathbf{D}^*}$. In this case, $m = 1$, and the 2-Hausdorff measure coincides with the measure defined by the hyperbolic 2-form, which on the upper half plane has the representation

$$\frac{dx \wedge dy}{y^2} \quad \text{if} \quad z = x + iy.$$

II, §4. NOGUCHI'S THEOREM ON D

In this section we give a theorem of Noguchi [No 3] in a general setting, using Noguchi's arguments.

Let $X \subset Y$ be a complex subspace of a complex space Y, hyperbolically imbedded. So far, we have looked at holomorphic maps of **D** into X. Now we shall look at maps of the punctured disc $\mathbf{D}^*$ into X, and we show that under certain circumstances, we get a result similar to that of Theorem 1.2, but with an added subtlety because of the puncture.

Theorem 4.1. *Let* $X \subset Y$ *be a relatively compact complex subspace, hyperbolically imbedded. Let* $\bar{X}$ *be the closure of* X *in* Y. *Let*

$$\bar{f}_n: \mathbf{D} \to \bar{X} \qquad and \qquad \bar{f}: \mathbf{D} \to \bar{X}$$

be a sequence of holomorphic maps, and a holomorphic map respectively. Let f_n *and* f *be their restrictions to* $\mathbf{D}^*$. *We suppose that* f_n, f *map* $\mathbf{D}^*$ *into* X *but* $f(0) \in \bar{X} - X$.

If $\{f_n\}$ *converges to* f *uniformly on compact subsets of* $\mathbf{D}^*$, *then* $\{\bar{f}_n\}$ *converges uniformly to* $\bar{f}$ *on compact subsets of* $\mathbf{D}$.

Proof. Let $\bar{f}(0) = y_0$. There is a neighborhood W of y_0 in Y such that W is a closed complex subspace of a polydisc, and the polydisc contains the ball $\mathbf{B}(y_0, 1)$ in some $\mathbf{C}^N$. In particular, W is complete hyperbolic. Suppose the conclusion of the theorem is false. We can then pick W such that for each positive integer k there are infinitely many n such that

$$\bar{f}_n(\mathbf{D}_{1/k}) \not\subset W.$$

For each $w \in \mathbf{B}(y_0, 1)$ and $r > 0$ we let $\mathbf{B}(w, r)$ be the open ball of center w and radius r, and we let $\mathbf{S}(w, r)$ be the sphere of center w and radius r. There is some k_0 such that

$$\bar{f}(\bar{\mathbf{D}}_{1/k_0}) \subset \mathbf{B}(y_0, \tfrac{1}{8}).$$

Since for each k, the sequence $\{f_n\}$ converges uniformly on the circle $\mathbf{S}_{1/k}$ (boundary of $\mathbf{D}_{1/k}$), there exists a subsequence $\{f_{n_k}\}$ of $\{f_n\}$ such that

(1) $f_{n_k}(\mathbf{S}_{1/k}) \subset \mathbf{B}(y_0, \tfrac{1}{4})$.

(2) $\bar{f}_{n_k}(\mathbf{D}_{1/k}) \not\subset \mathbf{B}(y_0, 1)$.

Hence for each $k \geq k_0$ there is a point $z_k \in \mathbf{D}_{1/k}$ such that

(3)   the point $x_k = \bar{f}_{n_k}(z_k)$ lies on $\mathbf{S}(y_0, \tfrac{1}{2})$.

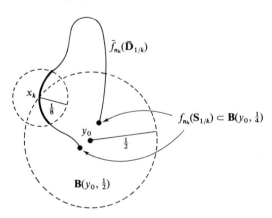

Let

$$E_k = \bar{f}_{n_k}(\mathbf{D}_{1/k}) \cap \mathbf{B}(x_k, \tfrac{1}{8}).$$

Then E_k is a 1-dimensional complex subspace of $B(x_k, \tfrac{1}{8})$. By Theorem 3.6 it follows that there is a constant C_0 (actually $\pi/64$ but who cares) such that the Euclidean measure of E_k satisfies

$$C_0 \leqq \mu^2_{\text{euc}}(E_k) \qquad \text{for all} \quad k = 1, 2, \ldots .$$

On the compact set $\bar{\mathbf{B}}(y_0, 1) \cap Y$, all length functions are equivalent, i.e. each is less than a constant times the other, so the corresponding distance functions are equivalent, and so are the corresponding Hausdorff measures. If H is a length function on Y, with distance d_H, we let:

μ^2_H = 2-dimensional Hausdorff measure defined by d_H, restricted to X;

μ^2_X = 2-dimensional Hausdorff measure defined by d_X on X.

Then we get

$$\mu^2_{\text{euc}}(E_k) \leqq C_1 \mu^2_H(E_k).$$

Since $d_H \ll d_X$ by the hypothesis that X is hyperbolically imbedded in Y, it follows that

$$\mu^2_H(E_k) \leqq C_2 \mu^2_X(E_k).$$

On the other hand, since f_{n_k} is distance decreasing from $d_{\mathbf{D}*}$ to d_X, it follows that

$$\mu^2_X(E_k) \leqq C_3 \mu^2_{\mathbf{D}*}(\mathbf{D}^*_{1/k}) \to 0 \qquad \text{as} \quad k \to \infty.$$

This succession of inequalities, starting with C_0, contradicts the fact that $\mu^2_{\text{euc}}(E_k)$ is bounded away from 0, and concludes the proof.

II, §5. THE KIERNAN–KOBAYASHI–KWACK (K^3) THEOREM AND NOGUCHI'S THEOREM

The theorem of this section evolved via [Kw], [Ko 4], VI, Theorems 6.1 and 6.2, and [Ki 3]. I follow [Ki 3].

To handle the higher dimensional case of the K^3 theorem, we need to start an induction in dimension 1, and then we need a slightly more general version of Kwack's theorem of §2, when we consider a sequence of maps $\{f_k\}$ instead of a single map f.

Lemma 5.1. *Let X be hyperbolically imbedded subspace of a complex space Y and assume X relatively compact. Let*

$$f_k: \mathbf{D}^* \to X$$

be a sequence of holomorphic maps. Let $\{z_k\}$ and $\{z'_k\}$ be sequences in $\mathbf{D}^*$ converging to 0 in $\mathbf{D}$, and such that

$$f_k(z'_k) \to y \in Y.$$

Then:

    (i) $f_k(z_k) \to y$ also.
    (ii) $f_k(0) \to y$.

Proof. Having assumed X relatively compact, we know from Theorem 2.1 that every holomorphic map of $\mathbf{D}^*$ into X extends to a holomorphic map of $\mathbf{D}$ into Y. In particular, $f_k(0)$ is defined by this extension. The proof of the present theorem is obtained by repeating the proof of Theorem 2.1, with the extra set of indices on f. Dealing with a single f was not needed in the proof.

Let M be a complex manifold and let A be a divisor. We say that A has **normal crossings** if at each point there exists a system of complex coordinates $z_1, \ldots, z_m$ for M such that locally

$$M - A = \mathbf{D}^{*r} \times \mathbf{D}^s \qquad \text{with} \quad r + s = m.$$

Thus A is defined locally by $z_1 \cdots z_r = 0$. We say that A has **simple normal crossings** if after expressing $A = \sum A_j$ as a sum of irreducible components, all A_j are non-singular (so there are no self-intersections) and A has normal crossings.

Theorem 5.2 (The K^3 theorem). *Let A be divisor with normal crossings on a complex manifold M. Let $X \subset Y$ be a relatively compact hyperbolically imbedded complex subspace. Then every holomorphic map*

$$f: M - A \to X$$

extends to a holomorphic map of M into Y.

Proof. By assumption, we may assume that

$$M = \mathbf{D}^m \qquad \text{and} \qquad M - A = \mathbf{D}^{*r} \times \mathbf{D}^s \qquad \text{with} \quad r + s = m.$$

The proof is by induction on $m = \dim M$. We do it in three steps.

    1. If $M - A = \mathbf{D}^*$ then the result is Kwack's theorem.

2.   Assume that we can extend f when $M - A = \mathbf{D}^{*n}$ for some n. We show that this implies we can extend f if $M - A = \mathbf{D}^{*n} \times \mathbf{D}^s$ for any s. Let

$$(w, t) = (w_1, \ldots, w_n, t_1, \ldots, t_s)$$

be the variables of $\mathbf{D}^n \times \mathbf{D}^s$. Let

$$f : \mathbf{D}^{*n} \times \mathbf{D}^s \to X$$

be holomorphic. For each t let $f_t(w) = f(w, t)$. Then by assumption we can extend f_t holomorphically to $\mathbf{D}^n$ for each t. It then suffices to prove that the map

$$(w, t) \to f(w, t)$$

is continuous. [This is a basic and elementary fact about functions of several variables. It is easily proved by writing down Cauchy's formula with respect to each variable, to get the power series expansion for f at a given point.]

Suppose f is not continuous at some point, say $(w, 0)$. Then there exists a sequence of points $\{(w^k, t^k)\}$ in $\mathbf{D}^{*n} \times \mathbf{D}^s$ converging to $(w, 0)$, and such that the sequence

$$f(w^k, t^k) \to y \neq f(w, 0).$$

Define

$$f_k : \mathbf{D}^* \to X \qquad \text{by} \qquad f_k(z) = f(w^k, z).$$

Since $t^k \to 0$ and $f_k(t^k) = f(w^k, t^k) \to y$, Lemma 5.1(ii) shows that

$$f_k(0) = f(w^k, 0) \to y.$$

But f_t is continuous for each t, and therefore

$$f(w^k, 0) \to f(w, 0) \neq y.$$

This contradicts the assumption that f is not continuous, and concludes the second step of the proof.

3.   Assume that f can be extended if $M - A = \mathbf{D}^{*n} \times \mathbf{D}^s$ for all s. We then show that f can be extended if $M - A = \mathbf{D}^{*n+1}$. By induction, f can be extended to $\mathbf{D}^{n+1} - \{(0, \ldots, 0)\}$. The map

$$g : \mathbf{D}^* \to X \qquad \text{defined by} \qquad g(z) = f(z, \ldots, z)$$

extends to $\mathbf{D}$. Define $f(0, \ldots, 0) = g(0)$. It suffices to prove that f is continuous.

If f is not continuous, there exists a sequence

$$(w^k, t^k) = (w_1^k, \ldots, w_n^k, t^k) \in \mathbf{D}^{*n+1}$$

such that

$$(w'', t'') \to (0, 0) \qquad \text{and} \qquad f(w^k, t^k) \to y \neq f(0, \ldots, 0).$$

By Lemma 5.1, with $f_k(z) = f(zw^k/|w^k|t_k)$ and $z_k' = |w^k|$, we see that

$$f(0, t^k) = f_k(0) \to y.$$

On the other hand, by Lemma 5.1 with

$$f_k(z) = f(zt^k/|t^k|, \ldots, zt^k/|t^k|, t^k) \qquad \text{and} \qquad z_k' = |t^k|,$$

we see that

$$f(0, t^k) \to f(0, \ldots, 0) \neq y.$$

This contradiction shows that f extends holomorphically, and proves the theorem.

The theorem was proved by Kwack when $X = Y$ is compact. In this case, A can be arbitrary, no assumption on the nature of the singularities is needed since one can extend the map to $M - \mathrm{Sing}(A)$, then to $M - \mathrm{Sing}(\mathrm{Sing}(A))$, and so forth. Kobayashi introduced the concept of "hyperbolically imbedded", and generalized her result to the case when X is hyperbolically imbedded in Y and A is non-singular. Kiernan treated the case when A has normal crossings, cf. [Ko 3], p. 93 as compared to [Ki 3].

I shall now give examples as in Kiernan [Ki 3].

Example. This example shows that if X is not compact, then restrictions on the singularities are necessary to make the conclusion valid. Let

$$X = \mathbf{D} \times (\mathbf{C} - \{1, -1\}) \subset \mathbf{P}^1(\mathbf{C}) \times \mathbf{P}^1(\mathbf{C}).$$

Since both $\mathbf{D}$ and $\mathbf{C} - \{1, -1\}$ are hyperbolically imbedded in $\mathbf{P}^1(\mathbf{C})$, it follows that X is hyperbolically imbedded in $\mathbf{P}^1(\mathbf{C}) \times \mathbf{P}^1(\mathbf{C})$. Let

$$M = \mathbf{D} \times \mathbf{D} \qquad \text{and} \qquad A = \{(z, w) \text{ such that } z = 0 \text{ or } z = \pm w\}.$$

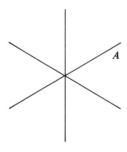

So A is the next worse thing to normal crossings. Define:

$$f: M - A \to X \qquad \text{by} \qquad f(z, w) = (z, w/z).$$

Then f does not extend to all of M since $f(0, 0)$ cannot be defined in a continuous way.

Kiernan also remarks that the above example is typical. Indeed, let A be a closed complex subspace of $M = \mathbf{D}^m$ with $(0, \ldots, 0) \in A$. Let L be the tautological bundle of lines over $\mathbf{P}^{m-1}(\mathbf{C})$. Then $M - A$ can be considered as a subspace of L. If $M - A$ is hyperbolically imbedded in L, then the inclusion $M - A \subset L$ is a counterexample to the singular version of the K^3 theorem.

As an immediate corollary of the resolution of singularities, however, Kiernan concludes:

Theorem 5.3. *Let A be a closed complex subspace of a complex space Z. Let X be relatively compact, hyperbolically imbedded in Y. Let*

$$f: Z - A \to X$$

be holomorphic. Then f extends to a meromorphic map of Z into Y.

This is just a consequence of the definitions of meromorphic map and desingularization.

We refer to Kiernan and Kobayashi for a discussion how these theorems apply to maps into moduli varieties, which are hyperbolically imbedded in their minimal (Satake) compactifications, according to Kobayashi-Ochiai [K-O 1].

Finally we give a version of Noguchi's theorem not in the context of sections, and in the higher dimensional case.

Theorem 5.4. *Let:*

$X \subset Y$ be a relatively compact, hyperbolically imbedded complex sub-space;

M be a complex manifold of dimension m;

A be a divisor with normal crossings in M.

Let

$$f_n: M - A \to X$$

be a sequence of holomorphic maps, which converges uniformly on compact subsets of $M - A$ to a holomorphic map

$$f: M - A \to X.$$

Let $\bar{f}_n, \bar{f}$ be their holomorphic extensions from M into Y.

Then in fact, $\{\bar{f}_n\}$ converges uniformly to $\bar{f}$ on every compact subset of M itself.

Proof. The proof is essentially the same as when $A = \mathbf{D}^*$. We merely carry an extra parameter along. But Lelong's Theorem 3.6 will still be used only in dimension 1.

The question of convergence arises in the neighborhood of a point $a \in A$. Without loss of generality, we may assume that complex coordinates $z_1, \ldots, z_m$ are chosen such that

$$M = \mathbf{D}^m, \qquad a = 0, \qquad \text{and } A \text{ is defined by } \quad z_1 \cdots z_p = 0.$$

Assume first that A is defined by $z_1 = 0$, so

$$M - A = \mathbf{D}^* \times \mathbf{D}^{m-1}.$$

Neighborhoods of 0 are given by

$$U_{k,r} = \mathbf{D}_{1/k} \times \mathbf{D}_r^{m-1} \qquad \text{with} \quad 0 < r < 1.$$

We let $\mathbf{S}_{1/k}$ as usual denote the circle of radius $1/k$. Let $\bar{f}(0) = y_0$. We let W denote a small open neighborhood of y_0 in Y. We can identify $W \subset \mathbf{B}(y_0, 1)$ with a complex subspace of a ball radius 1 in some $\mathbf{C}^N$. We write coordinates of $z \in \mathbf{D}^m$ as

$$z = (z_1, z') \qquad \text{with} \quad z_1 \in \mathbf{D} \qquad \text{and} \qquad z' \in \mathbf{D}^{m-1}.$$

If the sequence $\{\bar{f}_n\}$ does not converge uniformly on some neighborhood of 0, then we can pick W as above such that, given k, r there are infinitely many n for which

$$\bar{f}_n(U_{k,r}) \not\subset W.$$

For each $w \in \mathbf{B}(y_0, 1)$ and $t > 0$ we let $\mathbf{B}(w, t)$ be the open ball of radius t and center w, and we let $\mathbf{S}(w, t)$ be the sphere of center w and radius t. There exists k_0 and r_0 such that

$$\bar{f}(\bar{\mathbf{D}}_{1/k_0} \times \mathbf{D}_{r_0}^{m-1}) \subset \mathbf{B}(y_0, \tfrac{1}{8})$$

simply by the continuity of $\bar{f}$. Since $\{\bar{f}_n\}$ converges uniformly to $\bar{f}$ on $\mathbf{S}_{1/k} \times \bar{\mathbf{D}}_{r_0}^{m-1}$, there exists a subsequence $\{f_{n_k}\}$ and a sequence $\{z_k'\}$ of points $z_k' \in \mathbf{D}_{r_0}^{m-1}$ such that

$$\lim z_k' = 0$$

and

(1) $f_{n_k}(S_{1/k}, z'_k) \subset B(y_0, \frac{1}{4})$;

(2) $\bar{f}_{n_k}(D_{1/k}, z'_k) \not\subset B(y_0, 1)$.

Hence for each $k \geq k_0$ there is a point $z_{1k} \in D_{1/k}$ such that:

(3) the point $x_k = \bar{f}_{n_k}(z_{1k}, z'_k)$ lies on $S(y_0, \frac{1}{2})$.

Let

$$E_k = \bar{f}_{n_k}(D_{1/k}, z'_k) \cap B(x_k, \tfrac{1}{8}).$$

Then E_k is a 1-dimensional complex subspace of $B(x_k, \frac{1}{8})$. By Theorem 3.6 it follows that the euclidean measure of E_k satisfies

$$C_0 \leq \mu^2_{euc}(E_k) \qquad \text{for all} \quad k = 1, 2, \ldots.$$

On the compact set $\bar{B}(y_0, 1) \cap Y$ all length functions are equivalent, i.e. each is less than a constant times the other, so the corresponding distance functions are equivalent, and so are the corresponding Hausdorff measures. If H is a length function on Y, with distance d_H, we let:

$\mu^2_H = $ 2-dimensional Hausdorff measure defined by d_H, restricted to X;

$\mu^2_X = $ 2-dimensional Hausdorff measure defined by d_X on X.

Then we get

$$\mu^2_{euc}(E_k) \leq C_1 \mu^2_H(E_k).$$

Let $E^*_k = E_k - f_{n_k}(0, z'_k)$. Then $\mu^2_H(E_k) = \mu^2_H(E^*_k)$. Note that $E^*_k \subset X$. Since $d_H \ll d_X$ on X because X is hyperbolically imbedded in Y, it follows that

$$\mu^2_H(E_k) \leq C_2 \mu^2_X(E^*_k),$$

where μ^2_X is the Hausdorff measure associated with the Kobayashi distance d_X. On the other hand, since

$$f_{n_k} \colon M - A = D^* \times D^{m-1} \to X$$

is distance decreasing from d_{M-A} to d_X, it follows that

$$\mu^2_X(E^*_k) \leq C_3 \mu^2_{M-A}(D^*_{1/k}, z'_k).$$

But D^* is imbedded on (D^*, z'_k) in $M - A$, and the imbedding is Kobayashi distance decreasing. Therefore we obtain the final inequality

$$\mu^2_{M-A}(D^*_{1/k}, z'_k) \leq C_4 \mu^2_{D^*}(D^*_{1/k}) \to 0 \qquad \text{as} \quad k \to \infty.$$

This contradicts the first inequality in this successive chain, namely that $C_0 \leqq \mu_{euc}^2(E_k)$, and concludes the proof of the theorem in case

$$A = \mathbf{D}^* \times \mathbf{D}^{m-1}.$$

The case when $A = \mathbf{D}^{*p} \times \mathbf{D}^{m-p}$ is then done by induction, since the sequence $\{\bar{f}_n\}$ converges uniformly to $\bar{f}$ on compact subsets of

$$\mathbf{D}^{*p-1} \times \mathbf{D}^{m-p+1}.$$

This concludes the proof of the theorem.

The above theorem was formulated by Noguchi for sections of a proper family, see Theorem 3.2 in the next chapter. An analysis of the proof showed that the hypothesis of having sections was irrelevant, and that his arguments applied without change to arbitrary maps as formulated above.

[*Added in proofs*: I have received from Noguchi a preprint *Moduli Spaces of Holomorphic Mappings into Hyperbolically Imbedded Complex Spaces and Locally Symmetric Spaces*, in which Noguchi himself extends his theorem as in Theorem 5.4.]

CHAPTER III

Brody's Theorem

It followed immediately from the definition that every holomorphic map of $\mathbf{C}$ into a hyperbolic space is constant. Brody proved the converse under certain compactness conditions. This result has numerous applications, and reflects the global behavior of the space. This chapter is devoted to its proof and some of these applications.

III, §1. BOUNDS ON RADII OF DISCS

We suppose that X is a complex space with a length function. As in Chapter 0, §2 we define the norm with respect to this given length function. We let H denote the length function and let d_H be the associated distance.

Let $f: \mathbf{D} \to X$ be holomorphic. In line with Royden's approach (see Chapter IV, §1) we define

$$c_H(f) = c(f) = \sup_z |df(z)|,$$

where the sup is taken over all $z \in \mathbf{D}$ and may be ∞. We also define

$$c_H(X) = c(X) = \sup_f c(f) = \sup_f |df(0)|,$$

where the sup is taken over holomorphic maps $f: \mathbf{D} \to X$. Obviously $c(f) \leq c(X)$.

Lemma 1.1. *Let* $f: \mathbf{D} \to X$ *be holomorphic. Let* $z_1, z_2 \in \mathbf{D}$. *Then*

$$d_H(f(z_1), f(z_2)) \leq c(f) \, d_\mathbf{D}(z_1, z_2) \leq c(X) \, d_\mathbf{D}(z_1, z_2).$$

Proof. Let $\gamma: [0, 1] \to \mathbf{D}$ be a geodesic between z_1 and z_2 in $\mathbf{D}$. Then $f \circ \gamma$ is a curve joining $f(z_1)$ and $f(z_2)$, and its length is

$$\int_0^1 |df(\gamma(t))\gamma'(t)|_H \, dt \leq c(f) \int_0^1 |\gamma'(t)|_\mathbf{D} \, dt,$$

so

$$\text{length of } f \circ \gamma \leq c(f) \text{ length of } \gamma,$$

and the lemma follows.

Theorem 1.2 (Brody [Br]).

(i)   *If* $c(X)$ *is finite, then* X *is hyperbolic.*
(ii) *If* X *is compact hyperbolic, then* $c(X)$ *is finite.*

Proof. Suppose $c(X)$ finite. A Kobayashi chain gives rise to a Kobayashi path between two points. Applying Lemma 1.1 to each geodesic of a Kobayashi path, we conclude that for $x \neq y$ in X we have the inequality

$$d_X(x, y) \geq \frac{1}{c} \, d_H(x, y) > 0,$$

whence the first part of the thorem follows.

Conversely, suppose $c(X) = \infty$ and X is compact. There exists a sequence of holomorphic maps

$$f_n: \mathbf{D} \to X$$

with $|df_n(0)|$ increasing to ∞. By compactness, say $\lim f_n(0) = x$.

We now use the following basic lemma.

Lemma 1.3. *Let* X *be a complex space. Let* $f_n: \mathbf{D} \to X$ *be a sequence of holomorphic maps such that* $|df_n(0)| \to \infty$ *with respect to some length function on* X. *Suppose that* $f_n(0)$ *has a limit, and let* $x = \lim f_n(0)$. *Let* U *be an open chart at* x *in some* $\mathbf{C}^N$ *containing a closed ball* $\bar{\mathbf{B}}(x, s)$ *of some radius* $s > 0$ *with respect to the standard hermitian norm in the chart. Let* $\mathbf{S}$ *be the sphere* $\mathbf{S}(x, s)$. *Then there exists a point* $y \in \mathbf{S} \cap X$ *such that* $d_X(x, y) = 0$.

Proof. Let $r < 1$ be a positive number, and suppose that for an integer n we have

$$f_n(\bar{\mathbf{D}}_r) \subset \mathbf{B}(x, s).$$

Then

$$f_n'(0) = \frac{1}{2\pi\sqrt{-1}} \int_{C_r} \frac{f_n(w)}{w^2}\, dw, \qquad \text{so} \quad |f_n'(0)| \leq As/r,$$

where A is an appropriate constant bounding the length function in terms of the standard hermitian norm in the chart in $\bar{\mathbf{B}}(x, s)$. Since the hyperbolic metric at $z = 0$ is the same as the euclidean metric, it follows that if $|f_n'(0)|$ is large, then $r = r(n)$ has to tend to 0. Therefore given a positive integer m, there exists n such that $f_n(\mathbf{D}_{1/m})$ is not contained in U and therefore intersects the boundary $\mathbf{S}$. Hence we can find a point x_m on $\mathbf{S} \cap X$ such that $x_m = f_n(p_m)$ with $p_m \in \mathbf{D}_{1/m}$. Then

$$d_X(f_n(0), f_n(p_m)) \leq d_{\mathbf{D}}(0, p_m) \to 0$$

as $m \to \infty$. But a subsequence of $\{x_m\}$ converges to a point $y \in \mathbf{S} \cap X$. By the continuity of d_X we conclude that $d_X(x, y) = 0$, as was to be shown.

Actually the same argument proves the following statement:

Theorem 1.4. *Let X be a complex subspace of a complex space Y with a length function H.*

(i) *If $c(X)$ is finite, then X is hyperbolically imbedded in Y.*
(ii) *If X is relatively compact and hyperbolically imbedded in Y then $c(X)$ is finite.*

The proof is the same, using **H1 4**, which at the end of the proof gives the inequality

$$d_H(f_n(0), f_n(p_m)) \leq C d_X(f_n(0), f_n(p_m))$$

with a suitable positive constant C. The left hand side approaches s as $m, n \to \infty$, a contradiction which proves the theorem.

III, §2. BRODY'S CRITERION FOR HYPERBOLICITY

Let X be a complex space. We shall say that X is **Brody hyperbolic** if every holomorphic map $f: \mathbf{C} \to X$ is constant. Similarly, if X is a subset of a complex space Y, we say that X is **Brody hyperbolic in Y** (or **relative to Y**) if every holomorphic map $f: \mathbf{C} \to Y$ whose image is contained in X is constant. We usually omit the reference to Y if the context makes this reference clear.

Theorem 2.1. *Let X be a compact complex space. Then X is Brody hyperbolic if and only if X is Kobayashi hyperbolic.*

Since Kobayashi hyperbolic obviously implies Brody hyperbolic, the substance of Brody's theorem lies in the converse. Although the above statement is easy to remember, in applications, one must often have additional information, for instance when X is only relatively compact. Both of these features, which are given by Brody's proof, are contained in the next theorems, which will be referred to as **Brody's theorem**.

Theorem 2.2. *Let X be a relatively compact subset of a complex space Y. If $\bar{X}$ is Brody hyperbolic in Y, then there exists an open neighborhood of $\bar{X}$ in Y which is hyperbolic.*

Theorem 2.3. *Let X be a relatively compact complex subspace of a complex space Y. Suppose that X is not hyperbolically imbedded in Y (this is the case if X is not hyperbolic). Let H be a length function on Y. Then there exists a sequence of holomorphic maps*

$$g_n : \mathbf{D}_{r_n} \to X$$

defined on discs of increasing radii $r_n \to \infty$, and converging uniformly on every compact subset of $\mathbf{C}$ to a holomorphic map

$$g : \mathbf{C} \to \bar{X},$$

satisfying:

$$|dg_n(0)|_{\mathrm{euc}} = 1 = r_n \sup_{z \in \mathbf{D}_{r_n}} |dg_n(z)|,$$

$$|dg(0)|_{\mathrm{euc}} = 1 = \sup_{z \in \mathbf{C}} |dg(z)|_{\mathrm{euc}}.$$

Note: The norm for $dg_n(z)$ is with respect to the hyperbolic length function on $\mathbf{D}_{r_n}$, whereas the norm for $dg(0)$ is with respect to the standard euclidean norm on $\mathbf{C}$. We have used a subscript euc to remind the reader of these different lengths on the domain spaces. *In both cases, the length function is the given H on the image space.*

The proof will use Theorem 1.2. For this, we need a lemma which reparametrizes a holomorphic map

$$f : \mathbf{D}_r \to X$$

to another map

$$g : \mathbf{D}_r \to X$$

with a uniform bound on the derivatives of g. Observe that g will be obtained in such a way that the image of g is contained in the image of f.

Lemma 2.4 (Brody's reparametrization lemma). *Let X be a subset of a complex space with a length function. Let*

$$f: \mathbf{D}_r \to X$$

be holomorphic. Let c > 0, and for $0 \leq t \leq 1$ let

$$f_t(z) = f(tz).$$

(i) *If $|df(0)| > c$ then there exists $t < 1$ and an automorphism h of D such that if we put*

$$g = f_t \circ h$$

then

$$\sup_{z \in \mathbf{D}_r} |dg(z)| = |dg(0)| = c.$$

(ii) *If $|df(0)| = c$, then we get the same conclusion allowing $t \leq 1$.*

Proof. Let $m_t: \mathbf{D}_r \to \mathbf{D}_r$ be multiplication by t, so that f_t can be factored

$$\mathbf{D}_r \overset{m_t}{\to} \mathbf{D}_r \overset{f}{\to} X.$$

Then $dm_t(z)v = tv$ so

$$|df_t(z)| = |df(tz)| t \,\frac{1 - |z|^2/r^2}{1 - |tz|^2/r^2}.$$

Let

$$s(t) = \sup_{z \in \mathbf{D}_r} |df_t(z)|.$$

Note that if $t < 1$ then $|df_t(z)| \to 0$ for $|z| \to r$ so $|df_t(z)|$ has a maximum for $z \in \mathbf{D}_r$ and thus for $t < 1$,

$$s(t) = \max_{z \in \mathbf{D}_r} |df_t(z)|.$$

We have $s(0) = 0$. Also we can write $tz = w$. Taking the sup for $z \in \mathbf{D}_r$ amounts to taking the sup for $w \in t\mathbf{D}_r$. If $t < 1$, we can even take the sup over the closure $t\bar{D}_r$. It follows that $s(t)$ is continuous for

$$0 \leq t < 1.$$

Also $s(t) \to s(1)$ as $t \to 1$, even if $s(1) = \infty$. By assumption in the first part, $|df(0)| > c$, and hence $s(1) > c$. Hence there exists $0 \leq t < 1$ such that $s(t) = c$. Hence there is some $z_0 \in t\bar{\mathbf{D}}_r$ such that $|df_t(z_0)| = c$. Now let $h: \mathbf{D}_r \to \mathbf{D}_r$ be an automorphism such that $h(0) = z_0$ and let $g = f_t \circ h$. Then

$$|dg(0)| = |df_t(z_0)||dh(0)| = |df_t(z_0)| = c,$$

thus proving the first part. The second part is proved similarly, allowing $t = 1$. This concludes the proof of Lemma 2.4.

Remark. Although we won't need it, we observe, as Brody does, that $s(t)$ is increasing for $0 \leq t \leq 1$. Indeed, if $t_1 < t_2 < 1$, then there is some z_1 such that

$$s(t_1) = |df_{t_1}(z_1)|,$$

and we let $z_2 = t_1 z_1 / t_2$ so that $t_2 z_2 = t_1 z_1$. A simple comparison of the extra factor then shows that

$$|df_{t_1}(z_1)| < |df_{t_2}(z_2)| \leq s(t_2).$$

If $t < 1$ then the extra factor

$$t \, \frac{r^2 - |z|^2}{r^2 - |tz|^2} \quad \text{is less than 1,}$$

and if $t = 1$ then this extra factor is equal to 1, so $s(t)$ is also increasing up to $t = 1$.

Proofs of Theorems 2.2 and 2.3. These are proved simultaneously. Suppose first that every open neighborhood of $\bar{X}$ in Z is not hyperbolic. Let $\{U_n\}$ be a decreasing sequence of open neighborhoods of $\bar{X}$ such that

$$\bigcap U_n = \bar{X},$$

and such that $\bar{U}_n$ is compact for all n. By Theorem 1.2(i) there exists a sequence of holomorphic maps

$$f_n: \mathbf{D} \to U_n \quad \text{such that} \quad |df_n(0)| \to \infty.$$

After renumbering the sequence if necessary, and after making a dilation from $\mathbf{D}$ to $\mathbf{D}_{r_n}$ with suitable r_n, by Lemma 2.4 there exists

$$g_n: \mathbf{D}_{r_n} \to U_n \quad \text{such that} \quad |dg_n(0)|_{\text{euc}} = 1 = r_n \sup_{z \in \mathbf{D}_{r_n}} |dg_n(z)|_n,$$

and the radii r_n are increasing to infinity.

In order to avoid the double index, for the rest of this proof we write $\mathbf{D}_n$ instead of $\mathbf{D}_{r_n}$. Also we indexed the norm and wrote

$$|dg_n(z)|_n$$

to denote the hyperbolic norm on the disc of radius r_n, because the radii are now varying, and these norms are going to be compared with the euclidean norm on $\mathbf{C}$. At $z = 0$ the two norms differ by a factor of r_n.

For Theorem 2.3 the hypothesis and condition **HI 4** imply $c(X) = \infty$, so we can do as above, but with the maps f_n and g_n taking their values in X itself.

We want to show that a subsequence of $\{g_n\}$ converges uniformly on compact sets. By Ascoli's theorem, it will now suffice to prove that the sequence in equicontinuous on a given compact set K because the images of all g_n are contained in a relatively compact set. Note that

$$K \subset \mathbf{D}_n \qquad \text{for all sufficiently large } n.$$

Equicontinuity is a local property, so let $z_0 \in K$. Abbreviate $d_n = d_{\mathbf{D}_n}$. Then

$$d_H(g_n(z), g_n(z_0)) \leqq r_n d_n(z, z_0) \qquad \text{for} \quad z \in K,$$

because the distance is the inf of lengths of C^1 curves joining the two points, and curves in $\mathbf{D}_n$ map into curves of Z under g_n. For $n > m$, since $r_m > 1$ for m large, we get

$$r_n d_n(z, z_0) \leqq r_m d_m(z, z_0).$$

For fixed m, we deduce the equicontinuity for $n > n_0$, because on a fixed compact set K the distances $r_n d_n$ are equivalent to d_{euc}.

Thus finally, there is a subsequence of $\{g_n\}$ which converges uniformly on each compact subset of $\mathbf{C}$, and since the neighborhoods U_n shrink to $\bar{X}$, the limit mapping g must map $\mathbf{C}$ into $\bar{X}$. This limit mapping cannot be constant, because

$$|dg(0)| = 1,$$

so we have proved Theorems 2.2 and 2.3.

Remark. Although in Theorem 2.2 we assumed that X is a complex subspace (so locally closed) its closure $\bar{X}$ may not be a complex subspace. The boundary of X may be relatively messy. At this point it is quite essential that we defined a holomorphic map into $\bar{X}$ to be a holomorphic map into the ambient space, whose image is contained in $\bar{X}$.

Corollary 2.5. *Let X be a complex space, and assume that X is Brody hyperbolic. Let U be a connected open subset which is relatively compact. Then U is hyperbolically imbedded in X.*

Proof. Corollary of Theorem 2.3.

Theorem 2.6. *Let X be a complex space with a length function. Let $f: \mathbf{C} \to X$ be a holomorphic non-constant map, whose image is contained in a relatively compact S. Then there exists a non-constant holomorphic map $g: \mathbf{C} \to X$ whose image is contained in the closure $\bar{S}$, and such that*

$$|dg(0)| = \sup|dg(z)| = 1.$$

Proof. For each disc $\mathbf{D}_{r_n}$ with increasing radius r_n, we let f_n be the restriction of f to this disc. Then the same proof as above works. We observe that in Brody's reparametrization, the image of g_n is contained in the image of f_n, and so the image of the limit is contained in $\bar{S}$. This proves the theorem.

Remark. Suppose that the length function on X comes from a hermitian metric on some ambient complex manifold, and let ω be the associated positive $(1, 1)$-form. Then we see that

$$\int_{\mathbf{D}_r} g^*\omega \leq \pi r^2.$$

This is usually expressed by saying that g is of **order** ≤ 2, and in particular is of finite order.

III, §3. APPLICATIONS

As a first application, we recall that a map of topological spaces is called **proper** if it is closed, and if the inverse image of a compact set is compact.

Proposition 3.1. *Let $\pi: X \to Y$ be a proper holomorphic map of complex spaces.*

(i)   *If Y is hyperbolic and each fiber $\pi^{-1}(y)$ is hyperbolic for all $y \in Y$, then X is hyperbolic.*

(ii) *If there is some $y_0 \in Y$ such that $\pi^{-1}(y_0)$ is hyperbolic, then there exists a neighborhood U of y_0 in Y such that $\pi^{-1}(y)$ is hyperbolic for all $y \in U$.*

Proof. For (i), by Proposition 2.6 of Chapter I, it suffices to show that given $y \in Y$, there exists an open neighborhood U of y such that $\pi^{-1}(U)$ is hyperbolic. Take U open such that $\bar{U}$ is compact. Then $\pi^{-1}(U)$ is open and its closure is contained in $\pi^{-1}(\bar{U})$, and so is compact since π is proper. By Brody's Theorem 2.3, if $\pi^{-1}(U)$ is not hyperbolic, then there exists a non-constant holomorphic map $f: \mathbf{C} \to \pi^{-1}(\bar{U})$. Then $\pi \circ f$ is constant because Y is hyperbolic, so $f(\mathbf{C})$ is contained in a fiber, and is constant because all fibers are assumed hyperbolic. This concludes the proof for (i).

As for (ii), we use Theorem 2.2 that there is a hyperbolic neighborhood V of $\pi^{-1}(y_0)$. Then there is some open neighborhood U of y_0 such that $\pi^{-1}(U) \subset V$ (again using the fact that π is proper), so $\pi^{-1}(y)$ is hyperbolic for all $y \in U$, thus concluding the proof of the theorem.

We observe that the second part of the theorem is Brody's result that the property of being hyperbolic is an open condition in holomorphic families. It is not a closed condition, as was shown for the first time by Brody–Green [B–G], who gave the example of the family of hypersurfaces

$$x_0^d + \cdots + x_3^d + (tx_0 x_1)^{d/2} + (tx_0 x_2)^{d/2} = 0.$$

They prove that these varieties are hyperbolic for d even ≥ 50 and sufficiently general $t \neq 0$. But for $t = 0$ it is a Fermat hypersurface which contains lines, and so is not hyperbolic. These results answered questions of Kobayashi [Ko 4]. See Chapter VII, Theorem 4.2.

Let $\pi: X \to Y$ be a proper holomorphic map of complex spaces. Let $\mathrm{Sec}(\pi)$ be the set of holomorphic sections. Under the hypotheses of Proposition 3.1(i) these sections form a "normal" family, that is given any sequence of sections, there exists subsequence which converges uniformly on every compact subset of Y, by Ascoli's theorem as in Theorem 4.1 of Chapter I. We apply Ascoli to sections viewed as holomorphic maps

$$s_U: U \to \pi^{-1}(U),$$

where U is an open set in Y. If Y is compact, then there is a subsequence which converges uniformly on Y itself.

In applications to algebraic geometry, suppose that X and Y are imbedded in projective space. Let the degree refer to the projective degree. From the compactness of $\mathrm{Sec}(\pi)$, we then conclude that the set of sections has bounded degree.

This is a function-theoretic analogue of certain theorems or conjectures in number theory concerning bounded heights for rational points. The argument using Ascoli's theorem is originally due to Grauert [Gr], [Gr–R], developed by Riebesehl [Ri] . These authors use a definition of hyperbolicity based on differential forms and length functions (having to do with "negative curvature"). Their arguments are based on the distance decreasing properties of maps into "hyperbolic" spaces, and apply no matter what definition of hyperbolicity is adopted.

When the base space Y is compact, it is very rare that every fiber $\pi^{-1}(y)$ is hyperbolic. In his original paper [Gr], Grauert had to use other means from algebraic geometry to arrive at his proof of the Mordell conjecture for sections of algebraic families (bounding the degree of such sections), to take care of the finite number of possible degenerate fibers which are not hyperbolic. When the fibers have dimension 1, degenerate fibers may be rational lines, of genus 0. Thus the total space X is not hyperbolic around degenerate fibers.

However, Noguchi [No 3] has proved in this one-dimensional case that for a minimal model, it is possible to get a hyperbolic imbedding, so that the hypotheses of the following theorem are satisfied. By a **Riemann surface**, as usual, we mean a one-dimensional complex manifold.

Theorem 3.2 (Noguchi [No 3]). *Let $\bar{Y}$ be a Riemann surface and let*

$$\bar{\pi}: \bar{X} \to \bar{Y}$$

be a proper holomorphic map. Let S be a discrete subset of $\bar{Y}$ and let $Y = \bar{Y} - S$. Let $X = \bar{\pi}^{-1}(Y)$ and $\pi: X \to Y$ the restriction of $\bar{\pi}$ to X. Assume:

(a)   *For all $y \in Y$ the fiber $\pi^{-1}(y)$ is hyperbolic.*
(b)   *For each point $y \in S$ there is an open neighborhood U such that $\pi^{-1}(U - \{y\})$ is hyperbolically imbedded in $\bar{\pi}^{-1}(U)$.*

Then the set of sections $\mathrm{Sec}(\bar{\pi})$ is locally compact, that is every sequence of holomorphic sections has a subsequence which converges uniformly on compact subsets of $\bar{Y}$.

Proof. By Proposition 3.1, it suffices to consider the uniform convergence on some neighborhood of each point in $\bar{Y} - Y$. But this convergence is given by Theorem 4.1 of Chapter II, applied to sections

$$\bar{f}: U \to \bar{\pi}^{-1}(U),$$

where U is a neighborhood of a bad point y, such that $\bar{\pi}^{-1}(y)$ is not hyperbolic. We select U to be (isomorphic to) the disc $\mathbf{D}$, with y corresponding to the origin.

Let $\{f_n\}$ be a sequence of sections, defined over $U^* \approx \mathbf{D}^*$. Since $\pi^{-1}(U^*)$ is hyperbolic, and hyperbolically imbedded in $\bar{\pi}^{-1}(U)$, it follows from Kwack's Theorem 2.2 of Chapter II that f_n is the restriction of a holomorphic map $\bar{f}_n: U \to \bar{\pi}^{-1}(U)$. Similarly, after taking a subsequence,

$\{f_n\}$ converges to a section f over U^* which extends to a holomorphic section $\bar{f}$ over U. One can therefore apply Theorem 4.1 of Chapter II to complete the proof.

Furthermore, Noguchi has a version with a higher dimensional base, and a divisor with normal crossings on the base. We limited ourselves above to the case of a base of dimension 1 for simplicity, so that we would not have to give still another definition of hyperbolically imbedded in the context of fiber spaces, as Noguchi does.

The next theorems are applications of Brody's theorem due to Green [Gr 1], [Gr 2], [Gr 3]. We recall a definition. Let Y be a complex space. A **Cartier divisor** X on Y is a closed subspace which locally at every point can be defined by one analytic equation. This means: for each point $x \in X$ there is an open neighborhood V of x in Y such that $X \cap V$ is defined by one equation $\varphi = 0$, for some holomorphic function φ on V. For brevity, we omit the usual qualification "effective" since we don't deal with any other divisors in this section.

Also we recall that on a complex manifold, if X is a complex subspace of codimension 1 (all components have codimension 1) then X is a Cartier divisor. This is because the local ring of holomorphic functions at a point is factorial, and the irreducible divisors locally at the point are represented by the prime elements. In other words, their local equations are of the form $p = 0$, where p is a prime element in this local ring.

Theorem 3.3. *Let X be a Cartier divisor on a compact complex space Y with a length function.*

(i) *If $Y - X$ is not hyperbolically imbedded in Y, then there exists a holomorphic map $g : \mathbf{C} \to Y$ such that $g(\mathbf{C}) \subset X$ or $g(\mathbf{C}) \subset Y - X$, and*

$$|dg(0)| = 1 = \sup_z |dg(z)|.$$

(ii) *If X and $Y - X$ are Brody hyperbolic, then $Y - X$ is hyperbolically imbedded in Y and is complete hyperbolic.*

Proof. Suppose $Y - X$ is not hyperbolically imbedded in Y. By Theorem 2.3, there exists a sequence of holomorphic maps

$$g_n : \mathbf{D}_{r_n} \to Y - X$$

with $r_n \to \infty$ and a length function on Y such that

$$|dg_n(0)|_{\text{euc}} = 1 = r_n \sup_{z \in \mathbf{D}_{r_n}} |dg_n(z)|.$$

By a diagonal selection, after picking a subsequence we may assume without loss of generality that $\{g_n\}$ converges uniformly on compact subsets of $\mathbf{C}$ to a non-constant holomorphic map

$$g: \mathbf{C} \to (Y - X)^{\text{closure}}.$$

The problem is now whether the image of g can intersect X.

Lemma 3.4. *Let X be a closed complex subspace of a complex space Y. Let U be an open connected subset of $\mathbf{C}$. Let $g: U \to Y$ be holomorphic. Suppose that $g^{-1}(X)$ does not consist of isolated points. Then g maps U into X.*

Proof. Let $\{z_n\}$ be a sequence in $g^{-1}(X)$ converging to some number $z \in U$. Let V be an open neighborhood of $g(z)$ in Y on which X is defined by the finite number of analytic equations

$$\varphi_i = 0 \qquad \text{with} \quad i = 1, \ldots, m.$$

Let U_0 be a connected component of $g^{-1}(V)$ containing z_n for all n sufficiently large. Then $\varphi_i \circ g = 0$ on U_0, so g maps U_0 into X. Thus we have found some connected open set mapped into X by g. Since X is assumed closed in Y, it follows that for every open subset W of U with $g(W) \subset X$ we also have $g(\bar{W}) \subset X$. Furthermore, if $z \in \bar{W}$, then z is a limit point of elements of W, and so there exists an open neighborhood of z mapped into X by g. Now the open subsets of U mapping into X are inductively ordered by inclusion. A maximal element must be all of U by the preceding arguments, and the lemma is proved.

Lemma 3.5. *Let X be a Cartier divisor on a complex space Y. Let U be open connected in $\mathbf{C}$. Let*

$$g_n: U \to Y - X$$

be a sequence of holomorphic maps, converging uniformly to g. Then either g maps U into X or g maps U in $Y - X$.

Proof. By the preceding lemma, we may assume $g^{-1}(X)$ consists of isolated points. Suppose there is some $z_0 \in U$ such that $g(z_0) \in X$. Let $\mathbf{S}$ be a small circle centered at z_0 such that $g(\mathbf{S}) \subset Y - X$. Such a circle exists since $g^{-1}(X)$ consists of isolated points. We pick $\mathbf{S}$ small enough that $g(\mathbf{S}) \subset V$, where V is a small open set containing $g(z_0)$, in which X is defined by one equation $\varphi = 0$. Then the zeros of $\varphi \circ g$ are discrete on $g^{-1}(V)$, and we can pick $\mathbf{S}$ such that $\varphi \circ g$ has no zero on $\mathbf{S}$. We let $N(\varphi \circ g_n, \mathbf{S})$ be the number of zeros of $\varphi \circ g_n$ inside $\mathbf{S}$, so that

$$N(\varphi \circ g_n, \mathbf{S}) = \frac{1}{2\pi \sqrt{-1}} \int_{\mathbf{S}} d \log(\varphi \circ g_n)$$

and similarly with g instead of g_n if g has only isolated zeros, and no zeros on **S**. Then

$$\lim N(\varphi \circ g_n, \mathbf{S}) = N(\varphi \circ g, \mathbf{S})$$

because g_n maps the interior of **S** outside of X. It follows that $\varphi \circ g$ also has no zero inside **S**, contradicting $g(z_0) \in X$. This proves the lemma.

We have now proved (i), and also that if X and $Y - X$ are Brody hyperbolic, then $Y - X$ is hyperbolically imbedded in Y. In fact, we have proved something slightly stronger:

Let X be a Cartier divisor on a complex space Y. Let O be an open neighborhood of X in Y such that $O - X$ is relatively compact. If X and $\bar{O} - X$ are Brody hyperbolic then $O - X$ is hyperbolically imbedded in Y.

The proof is the same, getting a non-constant holomorphic map

$$g: \mathbf{C} \to (O - X)^{\text{closure}}.$$

Now to finish the proof of Theorem 3.3(ii). We know that $Y - X$ is locally complete hyperbolic in Y, as pointed out in the example preceding Theorem 1.4 of Chapter II. We then recall that d_{Y-X} defines the topology of $Y - X$ (because we have proved $Y - X$ hyperbolic), and that $Y - X$ is relatively complete in Y because $Y - X$ is relatively compact. We can then apply Theorem 1.3 of Chapter II to conclude the proof of Theorem 3.3.

For some applications, one cannot simply assume that X is hyperbolic in the above result, one has to assume something weaker. For instance, classical theorems that we recall below state that if one takes out sufficiently many hyperplanes in general position in projective space, then the complement is hyperbolic. But the hyperplanes contain complex lines. I give next a general formulation from [Gr 3].

Theorem 3.6. *Let X be a closed complex subspace of a compact complex space Y. Assume that X is the union of a finite number of Cartier divisors $X_1, \ldots, X_m$. Assume:*

(a)   *$Y - X$ is Brody hyperbolic;*
(b)   *for each choice of disjoint indices*

$$\{i_1, \ldots, i_k, j_1, \ldots, j_r\} = \{1, \ldots, m\}$$

the space $X_{i_1} \cap \cdots \cap X_{i_k} - (X_{j_1} \cup \cdots \cup X_{j_r})$ is Brody hyperbolic.

Then $Y - X$ is complete hyperbolic, and hyperbolically imbedded in Y.

Proof. This is proved in exactly the same way as Theorem 3.3. The only additional feature is the formal manipulation of indices to take into account the combinatorics of indices.

Example 1. Let $X_1, \ldots, X_m$ with $m \geq n + 1$ be hyperplanes in $\mathbf{P}^n$. We say that they are in **general position** if any $n + 1$ of them are linearly independent. It will be shown in Theorem 2.5 of Chapter VII that if $m \geq 2n + 1$ then such hyperplanes satisfy the hypotheses (a) and (b) of Theorem 3.6, and hence satisfy the conclusion.

Example 2. Given $n + 2$ hyperplanes in general position, we can define the **diagonals** as follows. After a linear change of coordinates, we can achieve that the hyperplanes are defined as follows. First $\mathbf{P}^n$ is imbedded in $\mathbf{P}^{n+1}$ as the hyperplane

$$x_0 + \cdots + x_{n+1} = 0,$$

where $x_0, \ldots, x_{n+1}$ are the homogeneous coordinates of $\mathbf{P}^{n+1}$. Then the $n + 2$ hyperplanes of $\mathbf{P}^n$ are the hyperplanes H_j defined by the equations

$$x_j = 0 \qquad \text{for} \quad j = 0, \ldots, n + 1.$$

The complement of these hyperplanes is the set

$$Y = \mathbf{P}^n - (H_0 \cup \cdots \cup H_{n+1}).$$

Let I be a subset of $\{0, \ldots, n + 1\}$ which consists of at least two elements and not more than n elements, say I. We let the corresponding **diagonal hyperplane** be:

$$D_I = \text{solutions of the equation } \sum_{i \in I} x_i = 0.$$

Note that if I' is the complement of I in the set of indices, then $D_I = D_{I'}$.

Let S be a complex subspace of a complex space Y. We shall say that Y is **Brody hyperbolic modulo** S if every holomorphic map $f : \mathbf{C} \to Y$ is either constant or contained in S.

The following theorem is due to Bloch for $n = 2$ [B] and Cartan for arbitrary n [Ca 1]. We shall prove it in Chapter VII.

Theorem 3.7. *Let Y be the complement of $n + 2$ hyperplanes of $\mathbf{P}^n$ in general position. Then every holomorphic map $\mathbf{C} \to Y$ is either constant or contained in the diagonals. In other words, Y is Brody hyperbolic modulo the diagonals.*

Green [Gr 1] and Fujimoto [Fu 2], [Fu 3] have given some complements to this theorem, which was then used and placed in the context of hyperbolicity by Kiernan–Kobayashi [K–K 2]. I use here a variation of "tautness" as in [K–K 2], namely the condition expressed in Theorem 1.4 of Chapter II to yield:

Theorem 3.8 (Cartan). *Let* $X = H_0 \cup \cdots \cup H_{n+1}$ *be the union of* $n + 2$ *hyperplanes in* $\mathbf{P}^n$, *in general position. Let* Δ *be the union of the diagonals. Then* $\mathbf{P}^n - X$ *is complete hyperbolic in* $\mathbf{P}^n$ *modulo* Δ, *and is hyperbolically imbedded in* $\mathbf{P}^n$ *modulo* Δ. *In fact,* $\mathrm{Hol}(\mathbf{D}, \mathbf{P}^n - X)$ *is relatively locally compact in* $\mathrm{Hol}(\mathbf{D}, \mathbf{P}^n)$ *mod* Δ.

The last statement will be proved in Chapter VIII. It implies the hyperbolic imbedding mod Δ by Theorem 1.4 of Chapter II, and the rest of the theorem then follows from Theorem 1.5 of Chapter II.

Example 3. We shall now give an example of Mark Green, exhibiting a Zariski open set which is Brody hyperbolic but not Kobayashi hyperbolic. We consider four lines in general position in $\mathbf{P}^2$. Let D_1, D_2, D_3 be the "diagonal" lines, drawn dotted on the figure.

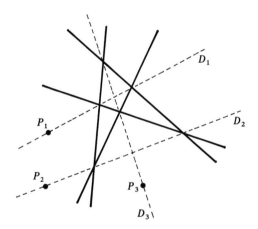

Let P_i ($i = 1, 2, 3$) be a point on D_i outside the four lines. Let

$$W = \mathbf{P}^2 - \{\text{the four lines}\} - \{P_1, P_2, P_3\}.$$

Then:

(1) W is Brody hyperbolic.
(2) W is not hyperbolic.

Proof. By the theorem of Bloch–Cartan, every holomorphic map

$$f : \mathbf{C} \to \mathbf{P}^2 - \{\text{the four lines}\}$$

must be constant or maps $\mathbf{C}$ into one of the diagonals D_1, D_2, D_3. But

$$D_i - P_i - (D_i \cap \text{the four lines}) \approx \mathbf{P}^1 - \{\text{three points}\},$$

which is hyperbolic, so every holomorphic map $\mathbf{C} \to W$ is constant, thus proving (1).

To prove (2), suppose the four lines are given by the equations

$$x_0 = 0, \qquad x_1 = 0, \qquad x_2 = 0, \qquad x_0 + x_1 + x_2 = 0$$

in the projective coordinates (x_0, x_1, x_2) of $\mathbf{P}^2$. Say $P_1 = (1, -1, 1)$. Let $r > 0$ and consider the map

$$f: \mathbf{D}_r \to \mathbf{P}^2 \qquad \text{given by} \qquad z \mapsto (1, e^z + h(z), -e^z),$$

where h is holomorphic, and chosen such that:

(a) $h(0) = h'(0) = 0$.
(b) $h(z) \neq 0$ when $e^z = -1$.
(c) $|h(z)| \leq \varepsilon e^{-r}$ for $z \in \mathbf{D}_r$ and ε small > 0.
(d) $h(1) = 0$.

We should actually write $f = f_r$, $h = h_r$ and $\varepsilon = \varepsilon_r$ depending on r. It is easy to see that $f: \mathbf{D}_r \to \mathbf{P}^2 - \{\text{four lines}\}$. Also P_1 is not in the image of f, and for small ε this image comes arbitrarily close to D_1, hence P_2 and P_3 also are not in the image of f. Let

$$x = f(0) = (1, 1, -1) \qquad \text{and} \qquad y = f(1) = (1, e, -e).$$

Hence

$$d_W(x, y) = d_W(f(0), f(1)) \leq d_{\mathbf{D}_r}(0, 1).$$

Letting $r \to \infty$ shows that $d_W(x, y) = 0$, thus proving (2). This concludes the proof of Green's example.

As Green himself observes, if one blows up the three points, one obtains the complement of a divisor with normal crossings which is Brody hyperbolic but not hyperbolic. However, Green's example is not affine, and I don't know an affine example. Is Brody hyperbolic equivalent to hyperbolic in the affine case?

Our next application will be a theorem of Kobayashi.

Let X be a compact complex manifold with tangent bundle TX. Let $\zeta: X \to TX$ be zero section. We say that the zero section can be **blown down to a point** if there exists a holomorphic map

$$f: TX \to Y$$

into a complex space Y such that f is a holomorphic isomorphism out-side $\zeta(X)$, and the image of $\zeta(X)$ under f is a point.

This condition at first seems strange, and after we have stated and proved Kobayashi's theorem, we shall make comments concerning this condition.

Theorem 3.9 (Kobayashi [Ko 5]). *Let X be a compact complex mani-fold such that the zero section of TX can be blown down to a point. Then X is hyperbolic.*

Proof (Urata [Ur]). There exists a holomorphic map

$$f: TX \to TX/\zeta(X) = Y,$$

where ζ is the zero section. Let $y \in Y$ be the image of the zero section. Let V be a hyperbolic neighborhood of y (for instance a polydisc in a chart). By Brody's theorem, it suffices to prove that X is Brody hyper-bolic. Let

$$g: \mathbf{C} \to X$$

be holomorphic. Then g lifts to the tangent bundle

$$g': \mathbf{C} \to TX,$$

and in fact, by the estimate for derivatives in Brody's theorem, we even know that $|g'(z)| \leqq 1$ for all z. After multiplying whatever the reader wants (either in $\mathbf{C}$, or the given length function) by a small constant, we may assume that $g'(\mathbf{C})$ lies in $T_c X$ for some small $c > 0$, namely the set of vectors v such that $|v| \leqq c$. We pick c small enough that $f(T_c X) \subset V$. Then

$$f \circ g': \mathbf{C} \to V$$

is holomorphic, so constant since V is hyperbolic. Since f is an isomorphism outside the zero section, it follows that $g' = 0$ and therefore g is constant. This proves the theorem.

We shall now comment on the condition in the theorem, and give an equivalent condition phrased in more ordinary terms. We place our-selves in the category of algebraic varieties.

Let E be a complex vector bundle over a complex manifold X. Then we can form the projective bundle of lines $\mathbf{P}(E) = \mathbf{P}E$, and the tautologi-cal line bundle $L = L_{\mathbf{P}E}$ on $\mathbf{P}E$ which at each point $[v] \in \mathbf{P}E$ associates the one-dimensional vector space L_v, namely the line through v on E. Let $\pi: \mathbf{P}E \to X$ be the projection.

In general, a line bundle L on a variety Y is said to be **ample** if there exists a positive integer m such that a basis $(s_0, \ldots, s_N)$ of the sections $H^0(Y, L^{\otimes m})$ gives a projective imbedding

$$\varphi_m \colon Y \to \mathbf{P}^N \qquad \text{by} \qquad y \mapsto (s_0(y), \ldots, s_N(y))$$

and the sections $s_0, \ldots, s_N$ have no common zero. We extend this definition to vector bundles. The dual of the exact sequence

$$0 \to L \to \pi^* E$$

is a surjection $\pi^* E^\vee \to L^\vee \to 0$ of the dual bundle on $L^\vee$. We define $E^\vee$ to be **ample** if $L^\vee$ is ample.

Observe that E and L are very closely related. Indeed, let E^* denote the open subset consisting of the non-zero vectors in E, and similarly let L^* be the set of non-zero vectors in L. Then we have a holomorphic isomorphism

$$E^* \to L^*$$

in the natural way.

Together with E and L we shall also have to work with the dual bundles $E^\vee$ and $L^\vee$. We let $\operatorname{Sym}^n(E^\vee)$ be the n-th symmetric power of $E^\vee$. If E is a vector space, so X is a point, then $\operatorname{Sym}^n(E^\vee)$ is the vector space of homogeneous polynomials of degree n in $E^\vee$, and thus may be identified with the space of homogeneous polynomial functions of degree n on E. Then for a vector space E, it is an elementary fact of algebraic geometry that there is a natural isomorphism

$$H^0(\mathbf{P}E, L^{\vee n}) \approx \operatorname{Sym}^n(E^\vee).$$

Globalizing over a complex manifold X, there is a natural isomorphism of sections

$$H^0(\mathbf{P}E, L^{\vee n}) \xrightarrow{\approx} H^0(X, \operatorname{Sym}^n(E^\vee)).$$

Now suppose that $(s_0, \ldots, s_N)$ is a basis of $H^0(\mathbf{P}E, L^{\vee n})$. Then we may view these sections as sections of $\operatorname{Sym}^n(E^\vee)$ under the above isomorphism. If E is a vector space, then $\operatorname{Sym}^n(E^\vee)$ is the set of homogeneous polynomial functions on E, and these functions can be evaluated. Passing to vector bundles, then we get a map

$$f \colon E \to \mathbf{A}^{N+1} \qquad \text{by} \qquad v \mapsto (s_0(v), \ldots, s_N(v))$$

from E into affine space $\mathbf{A}^{N+1}$. Observe that the zero section of E goes to the point $(0, \ldots, 0)$.

If finally $L^\vee$ is ample, then f is an isomorphism between E^* and its image. Thus we have proved:

Let X be a non-singular projective variety. If $T^\vee X$ is ample, then the zero section of TX can be blown down to a point.

This statement is essentially elementary, following directly from the definitions. **Grauert's criterion** is that the converse holds. Thus ampleness of the cotangent bundle is characterized by the possibility of blowing down the zero section of the tangent bundle. The general version in the context of schemes can be found in Grothendieck's EGA II, 8.9.1. For our purposes here, the more elementary direction suffices to give us Kobayashi's original statement:

Theorem 3.10. *Let X be a compact complex manifold whose cotangent bundle is ample. Then X is hyperbolic.*

It was Urata who showed that one could use directly the property of blowing down the zero section to give a simpler proof.

III, §4. FURTHER APPLICATIONS: COMPLEX TORI

The applications of Brody's theorem in this section are due to Mark Green. They also illustrate the use of the bound on derivatives.

We recall that a **complex torus T** is the quotient of $\mathbf{C}^n$ by a lattice. Motivated by diophantine problems, I had conjectured the following statement, proved by Green [Gr 3].

Theorem 4.1. *Let X be a closed complex subspace of a complex torus* **T**. *Then X is hyperbolic if and only if X does not contain a translated complex subtorus $\neq 0$.*

Proof. A complex subtorus in X would give a complex line $\mathbf{C} \to X$ which would contradict Brody's theorem.

Conversely, if X is not hyperbolic, then Brody's theorem yields a holomorphic map

$$f: \mathbf{C} \to X$$

such that $|df(z)| \leqq 1$ and $|df(0)| = 1$, where we take the representation

$$\mathbf{T} = \mathbf{C}^n/\Lambda,$$

with a lattice Λ, and the hermitian structure on **T** obtained from the ordinary hermitian structure on $\mathbf{C}^n$. Then f lifts to a holomorphic map into the universal covering space of **T** making the following diagram commutative:

$$(f_1,\ldots,f_n) \nearrow \begin{array}{c} \mathbf{C}^n \\ \downarrow \\ \mathbf{T} = \mathbf{C}^n/\Lambda \end{array}$$

$$\mathbf{C} \qquad f \searrow$$

and we have

$$\sum_{i=1}^{n} |f'_i|^2 \leqq 1.$$

Hence f'_i is constant for all i by Liouville's theorem, so f_i is linear. After a translation of X we may assume without loss of generality that $f_i(0) = 0$ for all i, and f is a one-parameter subgroup. Since X is assumed to be a complex analytic subspace, the complex analytic closure of the image of f is itself a complex analytic group, which is therefore a complex subtorus contained in X. This concludes the proof.

We have used an elementary lemma:

Let G be a group with a topology (not necessarily Hausdorff, since in the applications it may be a Zariski topology). Assume that given $x \in G$, translation by x is bicontinuous, and also that $x \mapsto x^{-1}$ is bicontinuous. Let H be an abstract subgroup of G. Then the closure $\bar{H}$ is a subgroup.

Proof. The closure of a subset S is the intersection of all closed subsets of G containing S. Since $x \mapsto x^{-1}$ is bicontinuous, it follows that $\bar{H}^{-1}$ is a closed set containing H^{-1}, so

$$\bar{H} = \mathrm{closure}(H) = \mathrm{closure}(H^{-1}) \subset \bar{H}^{-1}.$$

Applying the inverse map to this inclusion yields $\bar{H}^{-1} \subset \bar{H}$, so

$$\bar{H} = \bar{H}^{-1}.$$

Next let $h \in H$. Then $hH \subset \bar{H}$, so $H \subset h^{-1}\bar{H}$ which is closed, so $\bar{H} \subset h^{-1}\bar{H}$, whence finally $h\bar{H} \subset \bar{H}$. Therefore $H\bar{H} \subset \bar{H}$. Similarly $\bar{H}H \subset \bar{H}$. Finally let $h \in \bar{H}$. Then $hH \subset \bar{H}$ by what we have just proved, and hence $H \subset h^{-1}\bar{H}$ so $\bar{H} \subset h^{-1}\bar{H}$, so finally $h\bar{H} \subset \bar{H}$. So we have proved

$$\bar{H}\bar{H} \subset \bar{H}.$$

This concludes the proof of the lemma.

Note that we did not assume G to be a topological group, because in applications, the composition map $G \times G \to G$ is not continuous. Only translation by single elements is bicontinuous. The topology in the application is the analytic Zariski topology, for which the closed sets are the closed complex subspaces. The closure of a set is then the intersection of all closed complex subspaces containing the given set. The next result is also from [Gr 4].

Theorem 4.2. *Let* **T** *be a complex torus, imbedded in projective space. Let* X *be a hyperplane section of* **T** *(so in particular, a Cartier divisor on* **T***) which is Brody hyperbolic. Then* **T** $- X$ *is complete hyperbolic and hyperbolically imbedded in* **T***.*

Proof. By Theorem 2.2 we know that X is hyperbolic. The theorem then follows from Theorem 3.3, once we have shown that **T** $- X$ is Brody hyperbolic. Suppose **T** $- X$ is not Brody hyperbolic, and so is not hyperbolically imbedded. By Theorem 3.3(i) and the argument of Theorem 4.1, we see that there exists a translated one-parameter subgroup

$$g: \mathbf{C} \to \mathbf{T} - X.$$

We observe that $g(\mathbf{C})$ has to come arbitrarily close to X. For otherwise, there exists an open neighborhood W of X such that $g(\mathbf{C})$ is contained in the complement of W. But the complement of a closed neighborhood of a hyperplane in projective space is holomorphically isomorphic to a bounded open set in affine space, and therefore $g(\mathbf{C})$ would have to be constant. Thus the intersection

$$\overline{g(\mathbf{C})} \cap X$$

is not empty. We shall now see that this leads to a contradiction, resulting from the following theorem, which I had conjectured in the context of the theory of transcendental numbers [La 1], and which was proved by [Ax].

Theorem 4.3. *Let* **T** *be a projective complex torus, and let* X *be a hyperplane section. Let* $g: \mathbf{C} \to \mathbf{T}$ *be a one-parameter subgroup. Then* X *contains a translation of* $g(\mathbf{C})$ *or the intersection of* X *and* $g(\mathbf{C})$ *is not empty.*

Proof. The one-parameter group gives rise to a group of translations on **T**. We let

$$\alpha: \mathbf{C} \times \mathbf{T} \to \mathbf{T}$$

be the operation of **C** on **T** arising from this group, and we write it additively, that is

$$\alpha(z, x) = g(z) + x.$$

Suppose X does not contain a translation of $g(\mathbf{C})$. We then prove that given $r > 0$ and an open neighborhood V in X of a point $x_0 \in X$, the restriction

$$\alpha: \mathbf{D}_r \times V \to \mathbf{T}$$

contains an open neighborhood of x_0 in $\mathbf{T}$. First we note that this map has finite fibers. For otherwise, we can write a fixed element in infinitely many ways as $g(z) + x$, with $z \in \mathbf{D}_r$ and $x \in V$. This implies that there is an element $x_1 \in V$ and infinitely many $z \in \mathbf{D}_{2r}$ such that $g(z) + x_1$ also lies in V. Such z have a point of accumulation z_1 in $\bar{\mathbf{D}}_{2r}$ and by continuity, $g(z_1) + x_1$ lies in the closure of V in X. Hence the map

$$g_1: \mathbf{C} \to \mathbf{T} \qquad \text{such that} \qquad g_1(z) = g(z) + x_1$$

intersects X in infinitely many points, and those z such that $g_1(z)$ lies in X are not discrete, so $g_1(\mathbf{C}) \subset X$ by Lemma 3.4.

By Gunning-Rossi [Gu-R], Proposition 2, p. 161, we can find r small enough and V small enough so that $\bar{\alpha}: \bar{D}_r \times \bar{V} \to \mathbf{T}$ is proper. Then by [Gu-R] again, the Proper Mapping Theorem, Theorem 3, p. 162, we conclude that the image of α is a complex subspace of the same dimension as $\mathbf{T}$, thus proving that the image $\alpha(\mathbf{D}_r \times V)$ contains a neighborhood of x_0.

As we saw before Theorem 4.3, there is an element $x \in \overline{g(\mathbf{C})} \cap X$. By what we have proved above, there exists a neighborhood U of x such that $U \subset g(\mathbf{D}_r) + X$ for some r. By definition of the closure, U contains some point of $g(\mathbf{C})$, so there exists $z_1 \in \mathbf{C}$ such that $g(z_1) \in U$, and so

$$g(z_1) = g(z_2) + x_1$$

for some $x_1 \in X$, whence $x_1 = g(z_1 - z_2)$ lies in $g(\mathbf{C})$. This concludes the proof.

Connection with diophantine geometry. I once conjectured that an affine open subset of an abelian variety has only a finite number of integral points in every finitely generated domain over $\mathbf{Z}$. For rational points on projective varieties the conjecture is that hyperbolic is equivalent to mordellic. Cf. [La 3]. For an affine variety (i.e. the complement of a very ample divisor in a projective variety), there are a priori three properties which may define "hyperbolicity":

Brody hyperbolic $\Leftarrow$ Kobayashi hyperbolic
                   $\Leftarrow$ hyperbolically imbedded in the projective closure.

The strongest property conjecturally implies finiteness of integral points, and conversely, this finiteness conjecturally implies the weakest. The situation is related to Theorem 3.3 and Green's Example 3 in §3.

CHAPTER IV

Negative Curvature on Line Bundles

This chapter gives sufficient conditions for a complex manifold to be hyperbolic in terms of differential forms. The key word here is curvature, which I find very misleading since what is involved are invariants from linear algebra and how they are related to distance or measure decreasing properties of holomorphic maps. The historical terminology, as it evolved from the *real* case, constitutes a serious psychological impediment for a beginner to learn the complex theory, or at least for me when I was a beginner. Not too much can be done about this, since the terminology of "curvature" is too well established to be discarded. But I have tried to speak systematically of Chern or Ricci forms, and to avoid "curvature" terminology, except in the title of the section as a code word, to obviate linguistic problems.

The first section on the Royden function is included here for want of a better place, but readers can start immediately with the rest of the chapter, which is essentially self contained, if the main interest is directed toward the Chern–Ricci forms and the Schwarz lemma. This first section will not be used in any essential manner.

However, the differential geometric properties of the Royden function are by no means cleared up. Even though it is usually not smooth, when the manifold X is compact and hyperbolic, what are its "curvature" properties? For lack of smoothness, one may have to view it as a distribution in order to take $dd^c \log$. This is probably a substantial area for some research.

The key section of this chapter is §3, the Ahlfors–Schwarz lemma, which gives a measure of hyperbolicity in terms of differential forms. It is a problem due to Kobayashi to determine the extent to which a hyperbolic manifold (compact?) is characterized by the existence of certain differential forms or similar objects.

In our considerations of line bundles, it is also appropriate here to discuss Kobayashi's notion of measure hyperbolic, having to do with the canonical line bundle and equidimensional holomorphic maps $f: \mathbf{D}^n \to X$ where $n = \dim X$ instead of maps $\mathbf{D} \to X$ as before.

IV, §1. ROYDEN'S SEMI LENGTH FUNCTION

This section is entirely due to Royden.

We describe Royden's semi length function, of which the Kobayashi distance is the integrated form, [Ro 1], [Ro 2].

The situation is not yet completely cleared up for complex spaces as distinguished from complex manifolds.

For the most part, we do not make real use of the Royden function, so this section could be omitted without logical harm to the rest of this chapter or to the book. But Royden's function does throw light on the infinitesimal aspect of the Kobayashi distance, for instance Theorem 1.2 of Chapter III.

Let us start with a complex manifold X, so we have no problem about what its tangent bundle means. We wish to define a semi length function on TX which is maximal such that every holomorphic map

$$f: \mathbf{D} \to X$$

has a differential with norm

$$|df(z)| \leq 1.$$

The norm of the linear map $|df(z)|$ is taken with respect to the hyperbolic length function on $\mathbf{D}$ and the expected semi length function on TX. Thus it is clear how to define this semi length function, which is called the **Royden semi length function**, or **Royden function** for short. For $v \in T_x X$ we define

$$H_X(v) = |v|_X = \inf 1/R \quad \text{such that there exist a holomorphic}$$
$$\text{map } f: \mathbf{D} \to X \text{ with } f(0) = x \text{ and}$$
$$f'(0) = Rv, \text{ with } R > 0.$$

It is obvious that this function is absolutely homogeneous of degree 1, i.e. satisfies **LF 2** from Chapter 0, §1, and also is ≥ 0. The only question lies with its continuity properties. We shall prove that it is upper semi continuous.

Just as holomorphic maps are Kobayashi distance decreasing, they are Royden function decreasing. In other words:

If $f: X \to Y$ is holomorphic, then

$$df: TX \to TY$$

is Royden function decreasing, that is $|df(x)| \leqq 1$ for all $x \in X$.

This is immediate from the definition.

Furthermore on the disc $\mathbf{D}$ itself, the Royden function is the same as the hyperbolic metric.

This is clear from the Schwarz–Pick lemma.

Let X, Y be two complex manifolds. Then

$$H_{X \times Y}(v, w) = \max[H_X(v), H_Y(w)].$$

This is obvious from the definitions.

Let H be a length function on X. Let K be a compact subset. Then there exists a constant C_K such that for all $v \in T_x X$ and $x \in K$ we have

$$|v|_X \leqq C_K H(v).$$

Proof. First it suffices to prove the statement locally. Thus we can assume K is contained in some polydisc W. Also on a compact set, all length functions are equivalent, so by shrinking the polydisc somewhat we can assume that the length function H is H_W, i.e. is the hyperbolic metric. Then we use the fact that an inclusion is Royden function decreasing, to get $H_X \leqq H_W$, whence the assertion follows.

Our next goal is to prove the first basic theorem of Royden.

Theorem 1.1. *The Royden function is upper semi continuous.*

For this we need the following lemma. We let $n = \dim X$.

Lemma 1.2. *Let $\varphi: \mathbf{D}_R \to X$ be holomorphic, and $\varphi'(0) \neq 0$. Then for each $r < R$ there is some $s > 0$ and a holomorphic map*

$$f: \mathbf{D}_r \times \mathbf{D}_s^{n-1} \to X$$

*such that f is a local isomorphism at the origin, and the restriction of f
to $\mathbf{D}_r \times 0^{n-1}$ is φ.*

The proof of this lemma in the general case is quite complicated
[Ro 2], and it would be desirable to have a simpler one. For a more
general result, see [Siu]. Here we give Royden's simple proof in the case
when X is open in $\mathbf{C}^n$.

Choose coordinates in $\mathbf{C}^n$ such that $\varphi'(0) = e_1$, where $e_1 = (1, 0, \dots, 0)$
is the first unit vector. Define

$$g: \mathbf{D}_R \times \mathbf{C}^{n-1} \to \mathbf{C}^n$$

by

$$g_1(z_1, \dots, z_n) = \varphi_1(z_1),$$

$$g_k(z_1, \dots, z_n) = \varphi_k(z_1) + z_k \qquad \text{for} \quad k \geq 2.$$

Then g is holomorphic on $\mathbf{D}_R \times \mathbf{C}^{n-1}$ and biholomorphic locally at the
origin. The restriction of g to $\mathbf{D}_R \times 0^{n-1}$ is equal to φ. Let

$$Z = g^{-1}(X) \qquad \text{and} \qquad \mathbf{D}_{r,s} = \mathbf{D}_r \times \mathbf{D}_s^{n-1}.$$

For s decreasing, the family $\{\bar{\mathbf{D}}_{r,s}\}$ is a decreasing family of compact sets
whose intersection is contained in Z. Hence

$$\bar{\mathbf{D}}_{r,s} \subset Z \qquad \text{for some} \quad s > 0.$$

Then we can take f to be the restriction of g to $\mathbf{D}_{r,s}$ to conclude the
proof of Lemma 1.2 in case $X \subset \mathbf{C}^n$.

Now for the proof of Theorem 1.1. Given $v \in T_x X$, there is

$$\varphi: \mathbf{D}_R \to X \qquad \text{with} \qquad \varphi(0) = x, \qquad \varphi'(0) = v, \qquad \frac{1}{R} < |v|_X + \varepsilon.$$

We take f as in Lemma 1.2 with r very close to R. Then $f^*v = e_1$, the
first unit vector. Then

$$f_* e_1 = v = f'(0) = \varphi'(0).$$

From the formula for the Royden function on a direct product (the max
of the function on the factors) we see that

$$|f^*v|_{\mathbf{D}_{r,s}} = |e_1|_{\mathbf{D}_{r,s}} = \frac{1}{r} < \frac{1}{R} + \varepsilon.$$

Since f is a local isomorphism at the origin of $\mathbf{D}_{r,s}$ it follows that df is a local holomorphic isomorphism from a neighborhood of e_1 in $T(\mathbf{D}_{r,s})$ to a neighborhood of v in TX. In particular, there exists a neighborhood V of v in TX such that for $v_1 \in V$ with $v_1 = f_* u_1$ we have:

$$|v_1|_X = |f_* u_1|_X \le |u_1|_{\mathbf{D}_{r,s}}$$
$$\le |e_1|_{\mathbf{D}_{r,s}} + \varepsilon$$
$$\le |v|_X + 2\varepsilon.$$

This concludes the proof of Theorem 1.1.

Remark. The extension lemma is used only to prove the semi continuity, and will not be used again.

In the terminology of Chapter 0, §1, we have proved that the Royden function is a semi length function.

Proposition 1.3. *The Royden function is the largest semi length function such that every holomorphic map $f : \mathbf{D} \to X$ is semi length function decreasing.*

Proof. This is immediate from the definitions, and is the analogue for the Royden function of the similar statement for the Kobayashi semi distance.

Now that we have obtained a semi length function, we can define the length

$$L_X(\gamma) = L_{HX}(\gamma) = \int_a^b |\gamma'(t)|_X \, dt$$

of C^1 curves, and then of (piecewise C^1) paths, as we did in the case of length functions. We define the **Royden semi distance** by

$$d_X'(x, y) = \inf L_X(\gamma)$$

taken over all piecewise C^1 paths in X between x and y. This d_X' satisfies the triangle inequality, and is ≥ 0.

Just as in the case of length functions, we have

$$d_X'(x, y) = \inf L_X(\gamma),$$

where γ ranges only on C^1 curves between x and y. This is because locally the Royden function is bounded by an ordinary length function, so the argument smoothing out corners works in this case.

Theorem 1.4. *We have* $d'_X = d_X$, *that is the Royden semi distance is equal to the Kobayashi semi distance.*

Proof. That $d'_X \leq d_X$ follows from Chapter I, 1.3, that the Kobayashi semi distance is the largest semi distance for which holomorphic maps are distance decreasing.

Next the proof of the converse inequality $d_X \leq d'_X$ is a simplification by Royden of his original argument. Given two points x, y let $\gamma: [a, b] \to X$ be a C^1 curve between x, y such that

$$L_X(\gamma) = \int_\gamma H_X \leq d'_X(x, y) + \varepsilon.$$

Say γ is defined on $[0, 1]$. Since the Royden function is upper semi continuous, there exists a sequence of continuous functions $\{h_n(t)\}$ decreasing to the function $t \mapsto |\gamma'(t)|_X$. Hence there exists a continuous function $h(t)$ such that

$$|\gamma'(t)|_X \leq h(t),$$

but

$$\int_0^1 h(t)\, dt \leq \int_0^1 |\gamma'(t)|_X\, dt + \varepsilon \leq d'_X(x, y) + 2\varepsilon.$$

Also we find a partition $(t_0, \ldots, t_n)$ of the interval for γ such that for $s_i \in [t_{i-1}, t_i]$ we have an estimate for the Riemann sum

$$\sum_{i=1}^n h(s_i)(t_i - t_{i-1}) \leq d'_X(x, y) + 3\varepsilon.$$

Now we shall prove the following lemma, which is a sort of mean value theorem in the Royden–Kobayashi context.

Lemma 1.5. *Let* $\gamma: [a, b] \to X$ *be a parametrized C^1 curve. Then given* s, ε *there exists* δ *such that for all t in the interval with $|t - s| < \delta$ we have*

$$d_X\big(\gamma(t), \gamma(s)\big) \leq [|\gamma'(s)|_X + \varepsilon]|t - s|.$$

Proof. There exists a disc $\mathbf{D}_R$ such that

$$\frac{1}{R} < |\gamma'(s)|_X + \varepsilon,$$

and a holomorphic map $f: \mathbf{D}_R \to X$ such that

$$f(0) = \gamma(s) \qquad \text{and} \qquad f'(0) = \gamma'(s).$$

Let $W = \mathbf{D}_c^n$ be a coordinate polydisc centered at $\gamma(s)$, and let $W_{c/2}$ be the polydisc $\mathbf{D}_{c/2}^n$ of half the radius. For $x, y \in W_{c/2}$ we have

$$d_X(x, y) \leq d_W(x, y).$$

We can choose δ_1 such that $|t - s| < \delta_1$ implies $\gamma(t) \in W_{c/2}$.

Now f restricted to the real axis and γ are two parametrized curves, with the same tangent vector at $\gamma(s) = f(0)$. Hence there exists δ_2 such that for $|t - s| < \delta_2$ we have

$$d_W(\gamma(t), f(t - s)) \leq \varepsilon|t - s|,$$

because d_W defines the topology on the polydisc W. Then

(*) $$d_X(\gamma(t), f(t - s)) \leq \varepsilon|t - s|.$$

On the other hand, composing f with dilation $\mathbf{m}_R : \mathbf{D} \to \mathbf{D}_R$ we have directly from the definition of the Kobayashi distance, for some δ and $|t - s| < \delta$:

$$d_X(f(t - s), f(0)) \leq d_{\mathbf{D}_R}(t - s, 0)$$

$$\leq \left(\frac{1}{R} + \varepsilon\right)|t - s|$$

(**) $$\leq [|\gamma'(s)|_X + 2\varepsilon]|t - s|.$$

Therefore by the triangle inequality and (*), (**) we find

$$d_X(\gamma(t), \gamma(s)) \leq [|\gamma'(s)|_X + 3\varepsilon]|t - s|,$$

thereby proving Lemma 1.5.

Lemma 1.6. *Let* $\gamma : [a, b] \to X$ *be a* C^1 *curve, and suppose* h *is a continuous function on* $[a, b]$ *such that*

$$|\gamma'(t)|_X < h(t) \text{ for all } t.$$

Given $s \in [a, b]$ *there is an open interval* I_s *containing* s, *such that*

$$d_X(\gamma(t_1), \gamma(t_2)) \leq h(s)|t_2 - t_1|$$

for all $t_1, t_2 \in I_s$ *with* $t_1 \leq s \leq t_2$.

Proof. Let $\varepsilon = h(s) - |\gamma'(s)|_X$, and take $I_s = (s - \delta, s + \delta)$ where δ is the number in Lemma 1.5.

We may now return to the proof of the theorem.

From the compactness of the interval, taking the partition sufficiently small, we then get

$$d_X(x, y) \leq \sum d_X(\gamma(t_i), \gamma(t_{i-1}))$$

$$\leq \sum [h(s_i) + \varepsilon]|t_i - t_{i-1}| \qquad \text{by Lemma 1.6 or 1.5}$$

$$\leq d'_X(x, y) + 4\varepsilon \qquad \text{by the Riemann sum estimate.}$$

This proves Royden's theorem.

Remark for complex spaces. Grauert, Riebesehl, and Reckziegel in the singular case take the tangent space to be the closure of the tangent bundle over the regular elements, in the full Zariski tangent space, which is not a vector bundle.

One might use curves and paths which locally factor through a resolution of singularities. Then the extension theorem might be applied to the resolution to get the semi continuity in general.

IV, §2. CHERN AND RICCI FORMS

We begin by fixing some notation and terminology.

Let X be a complex manifold of dimension n, and let L be a holomorphic line bundle over X. As we deal only with holomorphic bundles, unless otherwise specified, we shall omit the word holomorphic to qualify them. Let $\{U_i\}$ be an open covering of X such that $L|U_i$ has a trivialization (holomorphic, according to our convention)

$$\varphi_i: L|U_i \to U_i \times \mathbf{C}.$$

Then

$$\varphi_{ij} = \varphi_i \circ \varphi_j^{-1}: (U_i \cap U_j) \times \mathbf{C} \to (U_i \cap U_j) \times \mathbf{C}$$

is an isomorphism, given by a holomorphic map

$$g_{ij}: U_i \cap U_j \to \mathbf{C}^* = \mathrm{GL}_1(\mathbf{C})$$

such that

$$\varphi_{ij}(x, z) = (x, g_{ij}(x)z).$$

Let s be a holomorphic section of L over X. Then s is represented by a holomorphic map

$$s_i: U_i \to \mathbf{C},$$

satisfying

$$s_i = g_{ij}s_j.$$

A length function H on L such that H^2 is smooth is called a **metric**. Suppose a covering family $\{(U_i, \varphi_{ij})\}$ represents L as above. Suppose given for each i a function (smooth)

$$\rho_i \colon U_i \to \mathbf{R}_{>0}$$

such that on $U_i \cap U_j$ we have

$$\rho_i = |g_{ij}|^2 \rho_j.$$

Then we say that the family of triplets $\{(U_i, \varphi_{ij}, \rho_i)\}$ **represents a metric**. It is clear how to define compatible families, or compatible triples in this context, and the metric itself is an equivalence class of such covering triples, or is the maximal family of compatible triples. We could also write a representative family as $\{(U_i, \varphi_i, \rho_i)\}$ using the isomorphisms φ_i instead of the transition functions φ_{ij}.

If s is a section of L over some open set U_i containing a point P, then we define

$$|s(P)|^2 = \frac{|s_i(P)|^2}{\rho_i(P)}.$$

The value on the right-hand side is independent of the choice of U_i, as one sees at once from the transformation law.

Instead of using indices i, if L is trivial over an open set U, so

$$L_U \approx U \times \mathbf{C},$$

we write $s_U \colon U \to \mathbf{C}$ for the map representing a section $s \colon U \to L$ over U, and then we also write

$$|s|^2 = |s_U|^2/\rho_U.$$

A metric as above will be denoted by ρ, for instance. Since at each point L is one-dimensional, we see that the metric is determined by a hermitian product in a trivial way. In particular, we obtain line bundles by considering the tangent or cotangent bundle of a one-dimensional complex manifold, often called a **Riemann surface**.

Observe that if we only assume that the square of the length function is smooth outside the zero section, and continuous on the zero section, then in fact it is smooth on the whole bundle.

This is obvious from the above local formula over U, and is a special feature of line bundles. It is a significant feature, because if we have for instance a holomorphic map

$$f \colon X \to Y$$

of complex manifolds, and if H is a length function on TY which is smooth outside the zero section of TY, and if dim $X = 1$, then the pull-back f^*H is an ordinary semi metric on X. This happens all the time taking $X = \mathbf{D}$ to be the unit disc. We shall eventually compare f^*H with the hyperbolic metric $H_\mathbf{D}$ in such cases.

We shall now associate some differential forms to a metric on a line bundle. First we review some terminology.

Let $z_1,\ldots,z_n$ be holomorphic coordinates for X over U. As usual, we have the operators ∂ and $\bar\partial$, where say for a function $f(z)$,

$$\partial f(z) = \sum_{k=1}^n \frac{\partial f}{\partial z_k}\, dz_k \quad \text{and} \quad \bar\partial f(z) = \sum_{k=1}^n \frac{\partial f}{\partial \bar z_k}\, d\bar z_k.$$

Then

$$d = \partial + \bar\partial.$$

The operators ∂, $\bar\partial$, and d extend to forms of arbitrary degree as usual, for instance

$$\bar\partial(f_{IJ}(z,\bar z)\, dz_I \wedge d\bar z_J) = \sum_{k=1}^n \frac{\partial f}{\partial \bar z_k}\, d\bar z_k \wedge dz_I \wedge d\bar z_J,$$

where $dz_I = dz_{i_1} \wedge \cdots \wedge dz_{i_p}$ and $d\bar z_J = d\bar z_{j_1} \wedge \cdots \wedge d\bar z_{j_q}$. A sum of terms

$$\omega = \sum_{\substack{|I|=p\\|J|=q}} f_{IJ}(z,\bar z)\, dz_I \wedge d\bar z_J$$

is called a form of type (p, q). The numbers p, q do not depend on the choice of holomorphic coordinates, because if g is holomorphic, then $\bar\partial g = 0$ by the Cauchy–Riemann equations.

We define the operator

$$d^c = \frac{1}{4\pi\sqrt{-1}}(\partial - \bar\partial).$$

The advantage of such an operator is that it is a real operator. If ω is a form such that $\omega = \bar\omega$, then $d^c\omega$ also satisfies this property. Note that

$$dd^c = \frac{\sqrt{-1}}{2\pi}\partial\bar\partial = \frac{1}{2\pi\sqrt{-1}}\bar\partial\partial.$$

Given a metric ρ on L we define the **Chern form** of the metric to be the unique form $c_1(\rho)$ such that on an open set U, in terms of the trivialization as above, we have

$$\boxed{c_1(\rho)|U = -dd^c \log |s|^2 = dd^c \log \rho_U}$$

for any holomorphic section s. The right-hand side is independent of the choice of holomorphic coordinates on U, because for any non-zero holomorphic function g (giving rise to a change of charts) we have

$$\partial\bar\partial \log |g|^2 = \partial\bar\partial \log(g\bar g) = 0,$$

so $dd^c \log(g\bar g) = 0$.

We say that a $(1, 1)$-form

$$\omega = \frac{\sqrt{-1}}{2\pi} \sum h_{ij}(z)\, dz_i \wedge d\bar z_j$$

is **positive** and we write $\omega > 0$, if the matrix $h = (h_{ij})$ is hermitian positive definite for all values of z. This condition is independent of the choice of holomorphic coordinates $z_1, \ldots, z_n$.

A metric ρ is called **positive** if $c_1(\rho)$ is positive.

Example. Projective space. Let $X = \mathbf{P}^n$ be projective n-space. Let $T_0, \ldots, T_n$ be the homogeneous variables, and let U_i be the open set of points such that $T_i \neq 0$. We let

$$z_0^{(i)} = T_0/T_i, \ldots, z_n^{(i)} = T_n/T_i \qquad \text{so} \quad z_i^{(i)} = 1.$$

Then $z_j^{(i)}$ with $j \neq i$ are complex coordinates on U_i. Then there is a line bundle L, called the **tautological line bundle**, whose transition functions are given by

$$g_{ij} = T_j/T_i.$$

Its sheaf of sections is usually denoted by $\mathcal{O}(1)$. Let us fix the index i, so we write simply

$$z = (z_0, \ldots, z_n) \qquad \text{where} \quad z_j = T_j/T_i.$$

The **tautological metric** ρ on L is defined on U_i by the function

$$\rho(z) = \rho_i(z) = \sum_{v=0}^{n} z_v \bar z_v.$$

Then its Chern form is computed using rules from freshman calculus for the derivative of a product and quotient, to give

$$c_1(\rho) = \frac{\sqrt{-1}}{2\pi} \partial\bar{\partial} \log \rho(z)$$

$$= \frac{\sqrt{-1}}{2\pi} \frac{1}{\rho(z)^2} \left(\sum h_{ij} \, dz_i \wedge d\bar{z}_j \right),$$

where $h = (h_{ij})$ is the matrix

$$h = \rho(z)I - (\bar{z}_i z_j).$$

The metric on the cotangent bundle defined by this Chern form is called the **Fubini–Study metric**.

Proposition 2.1. *The Fubini–Study metric is positive.*

Proof. We have to show that h is positive definite (it is obviously hermitian). For any complex vector $C = {}^t(c_1, \ldots, c_n)$ we expand ${}^t\bar{C}hC$ and the Schwarz inequality immediately shows that for $C \neq O$ we have

$$^t\bar{C}hC > 0.$$

This proves the proposition.

The example will play no role for the rest of this chapter, but becomes useful later.

We now pass to volume forms. In $\mathbf{C}^n$ we have what we call the **euclidean form**, expressed in terms of coordinates z by

$$\boxed{\Phi(z) = \prod_{i=1}^{n} \frac{\sqrt{-1}}{2\pi} dz_i \wedge d\bar{z}_i.}$$

Except for the normalizing factor involving π, it is just the usual

$$dx_1 \wedge dy_1 \wedge \cdots \wedge dx_n \wedge dy_n.$$

The product sign is to be interpreted as the alternating product, but 2-forms commute with all forms, so it is harmless to write it as the usual product sign to emphasize this commutativity.

By a **volume form** on X, we mean a form of type (n, n), which locally in terms of complex coordinates can be written as

$$\Psi(z) = h(z)\Phi(z),$$

where h is C^∞ and $h(z) > 0$ for all z. This is invariant under a change of complex coordinates, since the factor coming out in such a change is of the form $g(z)\overline{g(z)}$, where $g(z)$ is holomorphic invertible.

Thus a volume form is a metric on the **canonical bundle**

$$K_X = \bigwedge^{\text{top}} T^{\vee}(X),$$

which is the top (n-th) exterior power of the cotangent bundle.

We define the **Ricci form** of Ψ to be the Chern form of this metric, so $\text{Ric}(\Psi)$ is the *real* $(1, 1)$-form given by

$$\text{Ric}(\Psi) = c_1(\rho) = dd^c \log h(z) \quad \text{in terms of coordinates} \quad z.$$

Remarks. *If C is a constant then*

$$\text{Ric}(C\Psi) = \text{Ric}(\Psi).$$

If u is a positive smooth function, then

$$\text{Ric}(u\Psi) = \text{Ric}(\Psi) + dd^c \log u.$$

Both assertions are trivial from the definition.

A 2-form commutes with all forms. By the n-th power

$$\text{Ric}(\Psi)^n$$

we mean the n-th exterior power. Then $\text{Ric}(\Psi)^n$ is an (n, n)-form, and in particular a top degree form on X. Since Ψ is a volume form, there is a unique function G on X such that

$$\frac{1}{n!} \text{Ric}(\Psi)^n = G\Psi.$$

We may also write symbolically

$$G = \frac{1}{n!} \text{Ric}(\Psi)^n / \Psi.$$

Note that G is a real-valued function. We call G the **Griffiths function** associated with the original volume form Ψ. (Cf. Chapter V, §3 and also [Gr 3].) We denote it by

$$G_\Psi \quad \text{or} \quad G(\Psi).$$

Special case. *Let* $\dim X = 1$, *and let* z *be a complex coordinate. Then the Griffiths function is given by*

$$G_\Psi(z) = \frac{1}{h(z)} \frac{\partial^2 \log h(z)}{\partial z\, \partial \bar{z}}.$$

Proof. Immediate from the definitions.

The function $-G$ in dimension 1 is classically called the **Gauss curvature**.

Example. *Let* $X = \mathbf{D}_a$ *be the disc of radius* a *with the volume form*

$$\Psi_a(z) = \frac{2a^2}{(a^2 - |z|^2)^2} \frac{\sqrt{-1}}{2\pi} dz \wedge d\bar{z}.$$

Then

$$\mathrm{Ric}(\Psi_a) = \Psi_a \quad \text{and so} \quad G(\Psi_a) = 1.$$

Proof. Immediate from the definitions.

In classical terms, the Gauss curvature of the disc is -1. We put the factor $2a^2$ in the definition of Ψ_a so that we would come out with $G = 1$ and Gauss curvature -1 for the hyperbolic disc. We may call Ψ_a the **normalized hyperbolic form on the disc** $\mathbf{D}_a$. For the standard unit disc, we also write

$$\Psi_{\mathbf{D}} = \Psi_1.$$

Example. In the case when $G(\Psi)$ is constant and the Ricci form is positive, the manifold is called **Einsteinian**. A positive constant multiple of the Ricci form is then taken as defining a hermitian metric, which is called an **Einstein–Kähler metric**.

We list further properties of the Ricci form.

Functoriality 2.2. *Let* X, Y *have the same dimension. Let* $f: Y \to X$ *be a holomorphic mapping. Let* Ψ_X *be a volume form on* X. *Then*

$$\mathrm{Ric}(f^*\Psi_X) = f^* \mathrm{Ric}(\Psi_X)$$

wherever $f^*\Psi_X$ *is positive, that is where* f *is a local isomorphism.*

If Ψ_Y is a volume form on Y then there is a function $u \geq 0$ such that $f^*\Psi_X = u\Psi_Y$, and we have

$$\mathrm{Ric}(f^*\Psi_X) = \mathrm{Ric}(\Psi_Y) + dd^c \log u \qquad \text{wherever} \quad u \neq 0.$$

Both assertions are immediate from the functoriality of ∂ and c_1.

Second derivative test 2.3. *Let u be a real function > 0 on Y. Let $y_0 \in Y$ be a point such that $u(y_0)$ is a maximum. Then*

$$(dd^c \log u)(y_0) \leq 0.$$

Proof. Given a tangent vector v in the tangent space at y_0, there exists an imbedding $f: U \to Y$ of an open disc U centered at 0 in $\mathbf{C}$ and a tangent vector (complex number) w such that

$$f(0) = y_0 \qquad \text{and} \qquad df(0)w = v.$$

By pull-back, it suffices to prove the negativity of the form $dd^c \log(u \circ f)$ at 0. Hence without loss of generality, we may assume $X = U$. With respect to the complex coordinate $z = x + iy$, $dd^c \log u$ is represented on U by

$$\left(\frac{\partial^2}{\partial x^2} + \frac{\partial^2}{\partial y^2} \right)(\log u) \frac{1}{4\pi} \, dx \wedge dy,$$

which is ≤ 0 at a maximum for u by elementary calculus.

Lemma 2.4. *Let $\omega_1, \ldots, \omega_n$ be positive $(1, 1)$-forms on the complex manifold X of dimension n. Then*

$$\omega_1 \wedge \cdots \wedge \omega_n$$

is a volume form. Furthermore, if $\omega_i \leq \eta_i$ for $i = 1, \ldots, n$ then

$$\omega_1 \wedge \cdots \wedge \omega_n \leq \eta_1 \wedge \cdots \wedge \eta_n.$$

Proof. Left to the reader. There is an excellent self-contained treatment of positivity in Harvey-Knapp [H-K].

From the lemma, it follows that if ω is a positive $(1, 1)$-form then

$$\Omega = \frac{1}{n!} \omega^n$$

is a **volume form**, said to be **associated** with ω.

Also a positive $(1, 1)$-form has an associated hermitian metric. One has to decide how to normalize this. We follow the classical convention. If

$$\omega = \frac{\sqrt{-1}}{2\pi} \sum h_{ij} \, dz_i \wedge d\bar{z}_j,$$

then the **associated hermitian metric** is taken to be defined by the positive definite hermitian matrix $(1/\pi)(h_{ij})$.

A **pair**

$$\boxed{(X, \omega) \quad \text{or} \quad (X, H)}$$

consisting of a complex manifold and a *positive* $(1, 1)$-form, or the length function H of a hermitian metric, will be called a **hermitian manifold**. We shall deal with such pairs all the time, and it is useful to have a notation which avoids repeating that ω is a positive $(1, 1)$-form. If ever one deals with a hermitian form which is not positive, then the convention has of course to be adjusted to need.

IV, §3. THE AHLFORS–SCHWARZ LEMMA

We begin with a one-dimensional result.

Theorem 3.1 (Ahlfors–Schwarz lemma). *Let* (X, Ψ) *be a one-dimensional hermitian manifold. Let*

$$f: \mathbf{D}_a \to X$$

be a holomorphic map. Let Ψ_a *be the normalized hyperbolic form on* $\mathbf{D}_a$. *Assume that there exists a number* $B > 0$ *such that* $G_\Psi \geqq B$. *Then*

$$Bf^*\Psi \leqq \Psi_a.$$

Proof. Write $f^*\Psi = u\Psi_a$ with a function u. It suffices to prove that $u \leqq 1/B$. We take two steps.

First step. We reduce to the case when u has a maximum in $\mathbf{D}_a$. Let $0 < t < a$. Then $\Psi_t \to \Psi_a$ as $t \to a$. Let u_t be the function such that

$$f^*\Psi = u_t \Psi_t \quad \text{on } \mathbf{D}_t.$$

Then for each $z \in \mathbf{D}_a$, $u_t(z) \to u(z)$ as $t \to a$. Write

$$f^*\Psi = h \frac{\sqrt{-1}}{2\pi} dz \wedge d\bar{z} \quad \text{on } \mathbf{D}_a.$$

Then h is bounded on $\mathbf{D}_t^{\text{cl}}$ (closure of $\mathbf{D}_t$, we avoid a bar since complex conjugates now occur simultaneous), and

$$u_t(z) = h(z)(t^2 - |z|^2)/2t^2,$$

so $u_t(z) \to 0$ as $|z| \to t$. Hence u_t has a maximum in $\mathbf{D}_t$, and it suffices to prove the inequality of the theorem for u_t, as desired.

Second step. Assume $u(z_0)$ is a maximum for u in $\mathbf{D}_a$. If $u(z_0) = 0$ we are done, because $u = 0$. Suppose $u(z_0) \neq 0$. Then f restricts to a local isomorphism $f: U \to X$ on a neighborhood U of z_0. Then by hypothesis

$$\Psi_a + dd^c \log u = f^* \operatorname{Ric}(\Psi) \geqq Bf^*\Psi.$$

The function $\log u$ has a local maximum at z_0 and hence

$$(dd^c \log u)(z_0) \leqq 0$$

by the second derivative test 2.3. This proves that

$$\Psi_a(z_0) \geqq Bf^*\Psi(z_0),$$

and therefore $u(z_0) \leqq 1/B$. Since $u(z_0)$ is a maximum value, this also proves our theorem.

The first higher dimensional version of the Schwarz–Ahlfors lemma is apparently due to Grauert-Reckziegel [G–R] and Dinghas [Di]. Extensions and clarifications of the differential geometric conditions under which the result is true were then given by Chern [Ch], Griffiths [Gri 2], Kobayashi [Ko 1], [Ko 4], and Reckziegel [Re]. In [G–R], Grauert-Reckziegel deal with length functions, which is more general, and applies to singular spaces as in Riebesehl [Ri]. I have limited myself here to the smooth case.

Let

$$\omega = \sum h_{ji}(z) \frac{\sqrt{-1}}{2\pi} dz_i \wedge d\bar{z}_j$$

be a positive $(1, 1)$-form. The **associated hermitian length function** H_ω has been defined to be the length function from the hermitian metric defined by the positive definite matrix $h = (1/\pi)(h_{ij})$ locally. We let d_ω denote the associated distance. Trivially:

$$\omega \leq \eta \quad \Rightarrow \quad H_\omega \leq H_\eta \quad \Rightarrow \quad d_\omega \leq d_\eta.$$

The following theorem appears in [G–R] in the context of length functions, and in Kobayashi [Ko 4] in the context of his hyperbolic manifolds.

Theorem 3.2. *Let (X, ω) be a hermitian manifold. Assume that there is a constant B such that for every holomorphic map $f : \mathbf{D} \to X$ we have*

$$Bf^*\omega \leq \Psi_\mathbf{D}.$$

Then X is hyperbolic and we have the inequalities

$$\sqrt{\frac{B\pi}{2}}\, H_\omega \leq H_X \qquad and \qquad \sqrt{\frac{B\pi}{2}}\, d_\omega \leq d_X,$$

where H_X and d_X are the Royden function and Kobayashi distance respectively.

Proof. The inequality $\sqrt{\dfrac{B\pi}{2}}\, H_\omega \leq H_X$ is immediate from the definitions. Then $\sqrt{\dfrac{B\pi}{2}}\, d_\omega \leq d_X$ follows immediately. Going back to the definition of the Kobayashi distance, one can also see directly and immediately that $\sqrt{\dfrac{B\pi}{2}}\, d_\omega \leq d_X$ without passing via the Royden function, for those who have not taken in the section on that function. That X is hyperbolic then follows, since the Kobayashi distance is bounded from below by a hermitian distance.

The above theorem applies to the following situation.

Theorem 3.3. *Let (X, ω) be a hermitian manifold. Assume that there exists a number $B > 0$ such that for every imbedded complex submanifold Y of dimension 1, we have*

$$G(\omega|Y) \geq B.$$

Then X is hyperbolic, and for every holomorphic map $f: \mathbf{D}_a \to X$, we have

$$Bf^*\omega \leqq \Psi_{\mathbf{D}_a} \qquad \sqrt{B}\, H_{f^*\omega} \leqq H_{\mathbf{D}_a}, \qquad \sqrt{\frac{B\pi}{2}}\, d_\omega \leqq d_X.$$

Proof. Again we reduce the statement to the case when $f^*\omega = u\Psi_{\mathbf{D}_a}$ and u has a maximum $u(z_0)$ in $\mathbf{D}_a$. If $u(z_0) = 0$, we are done. If $u(z_0) \neq 0$ then f is a local isomorphism at z_0 and

$$f: U \to f(U) = Y$$

thus gives an isomorphism of a neighborhood of z_0 with a one-dimensional complex submanifold of X. We can now apply the hypothesis and Theorem 3.1 to complete the proof of the first inequality. The second inequality follows from the definitions, the third is then immediate, and X is hyperbolic because the Kobayashi distance is bounded below by a hermitian distance. This concludes the proof.

In [Ko 4], Kobayashi raises the problem whether the converse is true, namely:

Let X be a compact complex manifold, even algebraic. Suppose X is hyperbolic. Does there exist a positive (1, 1)-form satisfying the condition of Theorem 3.3?

This problem involving a (1, 1)-form is the stronger version of alternative converses to Theorem 3.3, in which the (1, 1)-form is replaced by a length function, originally introduced by Grauert-Reckziegel [G-R], or similar weaker objects. For the equidimensional analogue, see the next section.

Remark 1. It is clear from the inequalities of the theorem that it is unnatural to take $-G(\omega|Y)$ as the fundamental function, rather than G itself. That is one reason why I don't like the convention of "curvature", and neither did Griffiths, for that matter, since he was the first one to eliminate it more or less systematically, and just exhibit the formal structure of several differential linear invariants coming into the theory [Gri 1]. I do find it useful, however, to define a (1, 1)-form ω to be **strongly hyperbolic** if it is positive and if there exists $B > 0$ such that for every imbedded complex submanifold Y of dimension 1 we have

$$G(\omega|Y) \geqq B.$$

If we merely have $G(\omega|Y) > 0$ for all Y then I would call ω just **hyperbolic**. When X is compact, the two are equivalent by Theorem 3.4 of Chapter V. Thus Theorem 3.3 can be stated in the form:

If there exists a strongly hyperbolic form, then X is hyperbolic.

Kobayashi's problem lies with the converse.

If Kobayashi's problem has an affirmative solution, well and good. If not, then there is some obstruction, and presumably there is some cohomological theory relating the complex analysis, the differential geometry and the topology of the manifold to explain why the hyperbolic form does not exist.

A number of examples of hyperbolic manifolds have been given by Green [Gr 6] by constructing hyperbolic forms. Green also gives counterexamples in [Gr 5]. As far as I know, it is an unsolved problem to determine whether the Brody–Green perturbation of the Fermat hypersurface

$$x_0^d + x_1^d + x_2^d + x_3^d + (tx_0 x_1)^{d/2} + (tx_0 x_2)^{d/2} = 0$$

with d even ≥ 50 and sufficiently general $t \neq 0$ admits a hyperbolic form. This would be an interesting special case of the general Kobayashi problem. That such hypersurfaces are hyperbolic will be proved by Nevanlinna theory in Chapter VII.

Remark 2. If X is compact, one may ask whether the seemingly weaker condition $G(\omega|Y) > 0$ for all Y implies the existence of $B > 0$ such that $G(\omega|Y) \geq B$ for all Y in the theorem. We shall prove this in the next Chapter V, Theorem 3.4, by giving another interpretation for the Griffiths function (negative of Gauss curvature on one-dimensional submanifolds).

Indeed, so far we have defined the Griffiths functions only for a volume form, but we used it for an imbedded one-dimensional submanifold. It is natural to define the **Griffiths function** of a $(1, 1)$-form ω (rather than a volume form) at a point x to be

$$G(\omega)(x) = \inf_Y G(\omega|Y)(x),$$

where the inf is taken over all locally imbedded one-dimensional submanifolds containing x, for instance over all complex discs in a chart centered at x. This definition then does not make it clear that $G(\omega)$ has any smoothness properties. In Chapter V, Theorem 3.4, when we identify this function with another one, the smoothness will be clear.

Note: The function $G(\omega)$ is the *negative* of what is usually called the (ugh!) **holomorphic sectional curvature**.

IV, §4. THE EQUIDIMENSIONAL CASE

Differential geometry

Instead of $(1, 1)$-forms, one can look at (n, n)-forms, where $n = \dim X$. This equidimensional case of the Schwarz lemma evolved from Dinghas [Di] and Chern [Ch] to Kobayashi [Ko 1], see also Griffiths [Gri 2] and Kobayashi-Ochiai [K-O 2]. The final formulation as given here is due to Kobayashi, who associated the Ricci form directly to a volume form without passing through a hermitian metric, following a point of view which he apparently heard from Koszul in another context.

Instead of the disc, we consider a polydisc

$$\mathbf{D}_a^n = \mathbf{D}_a \times \cdots \times \mathbf{D}_a \quad \text{(product taken } n \text{ times)}.$$

One could also take different radii $a_1, \ldots, a_n$. We then have the **normalized hyperbolic volume form**

$$\Psi_a^{(n)}(z_1, \ldots, z_n) = \prod_{i=1}^{n} \Psi_a(z_i).$$

Then

$$\text{Ric}(\Psi_a^{(n)})(z) = \sum_{i=1}^{n} \text{Ric}(\Psi_a(z_i)),$$

and

$$\frac{1}{n!} \text{Ric}(\Psi_a^{(n)})^n = \Psi_a^{(n)}, \quad \text{so} \quad G(\Psi_a^{(n)}) = 1.$$

Theorem 4.1. *Let X be a complex manifold of dimension n. Let Ψ be a volume form on X such that $\text{Ric}(\Psi)$ is positive, and such that there exists $B > 0$ satisfying*

$$G(\Psi) \geq B.$$

(This condition is automatically satisfied if X is compact.) Then for all holomorphic maps $f: \mathbf{D}_a^{(n)} \to X$ we have

$$Bf^*\Psi \leq \Psi_a^{(n)}.$$

Proof. Let $0 < t < a$. By reproducing exactly the argument in Step 1 of Theorem 3.3 in the present case of volume forms, we reduce the proof to the case when

$$f^*\Psi = u\Psi_a^{(n)}$$

and $u \geq 0$ has a maximum at z_0 in $\mathbf{D}_a^n$. Again if this maximum is 0, we are done, so we may assume the maximum $\neq 0$. Then

$$f^* \operatorname{Ric}(\Psi) = dd^c \log u + \operatorname{Ric}(\Psi_a^{(n)}).$$

By the second derivative test 2.3, this yields

$$f^* \operatorname{Ric}(\Psi)(z_0) \leq \operatorname{Ric}(\Psi_a^{(n)})(z_0).$$

We take the n-th power of each side and divide by $n!$. By definition of the Griffiths function, we know that

$$B\Psi \leq \frac{1}{n!} \operatorname{Ric}(\Psi)^n.$$

This shows that

$$Bf^*\Psi(z_0) \leq \Psi_a^{(n)}(z_0).$$

Hence $u(z_0) \leq 1/B$, and the theorem follows.

As pointed out by Kodaira in an appendix to [K–O 2], the above result holds under more general conditions as follows.

Let X be a complex manifold of dimension n as before. By a **pseudo volume form** Ψ we shall mean a continuous (n, n)-form which is C^∞ outside a proper complex subspace, and which locally in terms of complex coordinates can be expressed as

$$\Psi(z) = |g(z)|^{2q}h(z)\Phi(z),$$

where:

q is some fixed rational number > 0;
g is holomorphic not identically zero;
h is C^∞ and > 0;
$\Phi(z)$ is the euclidean volume form as defined in §3.

This definition is a variation of other possible definitions which weaken the conditions of a volume form, and is adjusted for some later applications. Zeros are allowed on a proper analytic subset.

We can define $\operatorname{Ric}(\Psi)$ for a pseudo volume form just as we did for a volume form, by the formula

$$\operatorname{Ric}(\Psi) = dd^c \log h.$$

Since g is assumed holomorphic, $dd^c \log|g|^{2q} = 0$ wherever $g \neq 0$.

Theorem 4.2. *Let X be a complex manifold, and let Ψ be a pseudo volume form such that $\mathrm{Ric}(\Psi)$ is positive, and such that there exists a number $B > 0$ satisfying*

$$B\Psi \leqq \frac{1}{n!} \mathrm{Ric}(\Psi)^n.$$

Then for all holomorphic maps $f : \mathbf{D}_a^{(n)} \to X$ we have

$$Bf^*\Psi \leqq \Psi_a^{(n)}.$$

Proof. The proof is identical with the previous proof. The arguments are valid under the weaker assumptions.

Measure theory

In Chapters I through III we dealt with Kobayashi hyperbolicity. There is a top-dimensional version, which we now mention in connection with the previous theorems.

Let Ψ be a pseudo volume form on X. Then Ψ defines a positive functional on $C_c(X)$ (continuous functions with compact support) by

$$\varphi \mapsto \int_X \varphi \Psi.$$

Hence by elementary measure theory, there is a unique positive measure μ_Ψ such that for all $\varphi \in C_c(X)$ we have

$$\int_X \varphi \Psi = \int_X \varphi \, d\mu_\Psi.$$

Let Z, X be complex spaces. We take for granted that if $f : Z \to X$ is analytic, and U open in Z then $f(U)$ is Borel measurable in X, in fact equal to a countable union of analytic subspaces of X. Cf. [Gu-R]. We suppose $\dim X = n$ as before.

We assume that there is a countable sequence of analytic maps

$$f_i : \mathbf{D}^{(n)} \to X \qquad (i = 1, 2, \ldots)$$

whose images cover X. Let A be a Borel measurable subset of X, and consider sequences $f_i : \mathbf{D}^{(n)} \to X$ of holomorphic maps, and open sets U_i in $\mathbf{D}^{(n)}$ such that

$$A \subset \bigcup f_i(U_i).$$

We define the **Kobayashi measure** on X (that is, the Borel sets of X) by

$$\mu_X(A) = \inf \sum_{i=1}^{\infty} \mu_{\Psi^{(n)}}(U_i),$$

where the inf is taken over all sequences $\{f_i\}$ and $\{U_i\}$ prescribed above. It is an exercise in basic techniques of measure theory to show that μ_X is a measure.

It can be shown that if A is a measurable set in X and $f: X \to Y$ is holomorphic, then $f(A)$ is Borel measurable in Y. Furthermore, a regular measure satisfies the property that the measure of a set is the inf of the measures of the open sets containing it. Hence in the definition of the Kobayashi measure, instead of taking open sets U_i we could take measurable sets in $\mathbf{D}^{(n)}$.

Let $f: X \to Y$ be holomorphic. Let μ, ν be regular measures on X and Y respectively. We say that f is **measure decreasing** if

$$\nu(f(A)) \leq \mu(A) \quad \text{for all measurable } A.$$

Instead of measurable A it would suffice to take open sets U.

Example. Let X, Y be complex manifolds and let Ψ_X, Ψ_Y be pseudo volume forms on X, Y respectively. If

$$f^*\Psi_Y \leq \Psi_X$$

then f is measure decreasing for the associated measures. Indeed, the set of points $x \in X$ such that $df(x)$ is singular is an analytic subset S, and $f(S)$ has measure 0. On the open complement of S, one sees at once that f is measure decreasing, so f is measure decreasing. This can be applied to the cases of Theorems 4.1 and 4.2, after multiplying Ψ with a sufficiently small constant.

As for the Kobayashi distance, we have the following properties for the Kobayasi measure. We let our complex spaces have dimension n.

4.3. *If $X = \mathbf{D}^n$ then $\mu_X = \mu_\Psi$, where $\Psi = \Psi_1^{(n)}$.*

4.4(a) *Let $f: X \to Y$ be a holomorphic map between complex spaces of dimension n. Then f is Kobayashi measure decreasing.*

4.4(b) *If μ is a measure on X such that every holomorphic map $f: \mathbf{D}^n \to X$ is measure decreasing from $\mu_{\mathbf{D}^n}$ to μ, then $\mu \leq \mu_X$.*

These properties are immediate from the definitions.

We define X to be **measure hyperbolic** if $\mu_X(V) > 0$ for all non-empty open subsets V of X.

Theorem 4.5. *Let X be a complex manifold and let Ψ be a pseudo volume form on X. Assume that $\mathrm{Ric}(\Psi)$ is positive, and that there exists a constant $B > 0$ such that*

$$B\Psi \leq \frac{1}{n!}\,\mathrm{Ric}(\Psi)^n.$$

(If X is compact such B always exists.) Then X is measure hyperbolic.

Proof. This is a corollary of Theorem 4.2 since the Kobayashi measure is bounded from below by the measure associated with a pseudo volume form, which is necessarily positive.

Theorem 4.6 (Kobayashi [Ko 4]). *Let X be a complex space of dimension n. If X is hyperbolic, then X is measure hyperbolic.*

Proof. We need some lemma from measure theory, and especially Hausdorff measure.

Lemma 4.7 ([H–W], p. 104). *Let X be a metric space with distance d, such that the Hausdorff measure satisfies*

$$\mu_d^{p+1}(X) = 0.$$

Let $x_0 \in X$. Then for all positive real numbers r except on a set of Lebesgue measure 0, we have

$$\mu_d^p(S(x_0, r)) = 0.$$

Proof. We start with a remark.

Let E be any measurable set. We let $\mathrm{diam}(E)$ denote the diameter of E, that is

$$\mathrm{diam}(E) = \sup d(x, y) \qquad \text{for} \quad x, y \in E.$$

Now let

$$d_1 = \inf_{x \in E} d(x_0, x) \qquad \text{and} \qquad d_2 = \sup_{x \in E} d(x_0, x).$$

Then $d_2 - d_1 \leq \mathrm{diam}(E)$. Furthermore

$$\int_0^\infty \mathrm{diam}(S(x_0, r) \cap E)^p \, dr = \int_{d_1}^{d_2} \mathrm{diam}(S(x_0, r) \cap E)^p \, dr$$

$$\leq \mathrm{diam}(E)^p \int_{d_1}^{d_2} dr \leq \mathrm{diam}(E)^{p+1}.$$

Coming to the proper part of the proof, by hypothesis for each positive integer n there is a sequence of balls $\{B_{n,i}\}$ such that

$$\lim_{n \to \infty} \sum_{i=1}^{\infty} r_{n,i}^{p+1} = 0, \quad \text{and} \quad X \subset \bigcup_{i=1}^{\infty} B_{n,i};$$

and also $r_{n,i} \leq c_n \to 0$ as $n \to \infty$. By the above inequality,

$$\sum_{i=1}^{\infty} \int_0^{\infty} \text{diam}(S(x_0, r) \cap B_{n,i})^p \, dr \leq \sum_{i=1}^{\infty} \text{diam}(B_{n,i})^{p+1}$$

$$\to 0 \quad \text{as} \quad n \to \infty.$$

On the other hand, in the left-hand side we can interchange sum and integral (say by Fubini's theorem, viewing the sum as an integral over the positive integers with the counting measure), because the integrand is non-negative. This means that if we put

$$f_n(r) = \sum_{i=1}^{\infty} \text{diam}(S(x_0, r) \cap B_{n,i}))^p,$$

then

$$\lim_{n \to \infty} \int_0^{\infty} f_n(r) \, dr = 0.$$

Since $f_n \geq 0$, it follows from standard Lebesgue integration that after subsequencing

$$\lim_{n \to \infty} f_n(r) = 0 \quad \text{for Lebesgue almost all } r.$$

By definition, this implies that $\mu_d^p(S(x_0, r)) = 0$ for almost all r, and concludes the proof of the lemma.

Hurewicz–Wallman give the above proposition in the context of "dimension theory". For our purposes we do not need to know anything about dimension theory. In our applications, we deal with complex spaces, whose regular points have the natural dimension, and which have a filtration by subspaces of decreasing complex dimension due to the singularities. Thus the concept of "dimension" is clear in the context of complex spaces. Formally, by induction, the lemma yields:

Proposition 4.8. *Let X be a metric space with distance function d, and "dimension" p (real dimension, that is), with $0 \leq p < \infty$. Then*

$$\mu_d^p(X) > 0.$$

In the application, we start with the image of a disc under a holomorphic map, which has "dimension" 2. We apply the lemma to this space, and conclude that if the 2-dimensional Hausdorff measure is 0, then the 1-dimensional Hausdorff measure of almost all circles around any point is 0. Then we apply the lemma again to each such circle, and land in a contradiction, since the 0-dimensional Hausdorff measure is just the counting measure on finite sets.

By a **projective variety** I shall mean an irreducible algebraic set in projective space. I end this section with

Conjecture 4.9. *Let X be a projective variety. Then X is hyperbolic if and only if all subvarieties of X (including X itself) are measure hyperbolic.*

The implication in one direction comes from Kobayashi's theorem, and from Brody's characterization of hyperbolic varieties by the condition that a holomorphic map $\mathbf{C} \to X$ is constant, so a subvariety of a hyperbolic variety is hyperbolic. The content of the conjecture lies in the converse: if all subvarieties are measure hyperbolic, then X is hyperbolic.

In connection with algebraic geometry, it is unknown if the property of being measure hyperbolic is birationally invariant. To deal with this problem, Yau [Yau] uses meromorphic maps instead of holomorphic maps to define a variant of the Kobayashi measure. He then has to prove the functorial properties concerning pull-backs, but once this is done, his definition makes the birational invariance obvious.

Existence of hyperbolic volume forms

A volume form Ψ on a complex manifold X will be called **strongly hyperbolic** if there exists $B > 0$ such that $\mathrm{Ric}(\Psi)^n/n! \geqq B\Psi$. We are concerned here with the existence of such forms. After multiplying Ψ with a suitable constant, one can always normalize B to be 1, so we get Ψ such that $G(\Psi) \geqq 1$, or in other words

$$\frac{1}{n!} \mathrm{Ric}(\Psi)^n \geqq \Psi.$$

We suppose that X is a compact complex manifold. Let L be a holomorphic line bundle on X. Let ρ be a metric on L. We have defined the **Chern form** $c_1(\rho)$. We say that $c_1(L) > 0$ if there exists a metric ρ on L such that $c_1(\rho) > 0$.

Let K be the canonical bundle. Suppose that $c_1(K) > 0$, so there exists a metric ρ on K such that $c_1(\rho) > 0$. This metric corresponds to a volume form Ψ, and by definition,

$$c_1(\rho) = \mathrm{Ric}(\Psi).$$

Since we assumed X compact, it follows that there exists $B > 0$ such that $G(\Psi) \geq B$, which achieves what we want.

More generally, let D be an effective divisor on X, expressed as a sum of irreducible divisors

$$D = \sum_{j=1}^{N} D_j.$$

Each D_j has an associated line bundle, which we denote by L_j. If ρ_j are metrics on L_j then we can form the tensor product

$$\rho_1 \otimes \cdots \otimes \rho_N \quad \text{on} \quad L_1 \otimes \cdots \otimes L_N.$$

We abbreviate the notation

$$X - D = X - \text{supp}(D) = X - |D|,$$

for the complement of the support of D in X.

Theorem 4.10 (Carlson-Griffiths [Ca–G], see also [Gri 3]). *Let X be a compact complex manifold. Let $D = \sum D_j$ be a divisor on X with simple normal crossings. Let L_D be the line bundle associated with D. If $c_1(K \otimes L_D) > 0$, then there exists a volume form Ψ on $X - D$ such that $G(\Psi) \geq 1$, in other words there exists a strongly hyperbolic volume form on $X - D$.*

Proof. Let ρ_K and ρ_j $(j = 1, \ldots, N)$ be metrics on K and L_j respectively. We put

$$\rho = \rho_K \otimes \rho_1 \otimes \cdots \otimes \rho_N.$$

By hypothesis we can choose these metrics such that

$$c_1(\rho) = c_1(\rho_K) + \sum_{j=1}^{N} c_1(\rho_j) > 0.$$

Let s_j be a holomorphic section of L_j whose divisor is D_j. Let $\alpha > 0$ be a positive real number. Let Ψ_K be the volume form associated with the metric ρ_K. Let

$$g = \prod_{j=1}^{N} |s_j|^2 (\log|\alpha s_j|^2)^2 \quad \text{and} \quad \Psi_\alpha = \Psi_K / g.$$

Thus g vanishes along the divisor and Ψ_α blows up. It will now suffice to prove the following statement, which is stronger than the one asserted in the theorem.

Lemma 4.11. *Given ε there exists $c > 0$ such that if $0 < \alpha \leq c$, then the form Ψ_α above satisfies:*

$$\mathrm{Ric}(\Psi_\alpha) = (1 - \varepsilon)c_1(\rho) + \psi,$$

$$\psi > 0 \quad and \quad \frac{1}{n!}\psi^n \geq B\Psi_\alpha \quad for \ some \quad B > 0.$$

Proof. We recall that locally, $c_1(\rho_j)$ is given by

$$c_1(\rho_j) = -dd^c \log|s_j|^2.$$

By definition

$$\mathrm{Ric}(\Psi_\alpha) = c_1(\rho) - \sum_{j=1}^{N} dd^c \log(\log|\alpha s_j|^2)^2$$

$$= (1 - 2\varepsilon)c_1(\rho) + \psi,$$

where

$$\psi = 2\varepsilon c_1(\rho) - \sum_{j=1}^{N} dd^c \log(\log|\alpha s_j|^2)^2.$$

Then

$$dd^c \log(\log|\alpha s_j|^2)^2 = \frac{2dd^c \log|\alpha s_j|^2}{\log|\alpha s_j|^2} - \frac{\sqrt{-1}}{\pi}\frac{\partial \log|\alpha s_j|^2 \wedge \bar{\partial} \log|\alpha s_j|^2}{(\log|\alpha s_j|^2)^2}.$$

The first term on the right is $-2c_1(\rho_j)/\log|\alpha s_j|^2$, which tends to 0 as $\alpha \to 0$, and may thus be absorbed in $\varepsilon c_1(\rho)$. Thus given ε, there is a positive (1,1)-form ω on X (actually $\varepsilon c_1(\rho)$, but it does not matter) such that for all α sufficiently small

$$\psi \geq \omega + \sum_{j=1}^{N} \frac{\sqrt{-1}}{\pi}\frac{\partial \log|\alpha s_j|^2 \wedge \bar{\partial} \log|\alpha s_j|^2}{(\log|\alpha s_j|^2)^2} > 0,$$

since each term in the sum is non-negative.

We must now prove that for α sufficiently small, ψ^n satisfies the desired inequality. By the compactness of X, it will suffice to do so in a neighborhood of a given point. If the point does not lie in D, then the inequality is obvious because locally any two C^∞ positive forms are bounded by a positive constant times each other. So suppose $x_0 \in D$. There is a system of complex coordinates $z_1, \ldots, z_n$ on an open neighborhood U of x_0 such that D_j is defined by $z_j = 0$ for $j = 1, \ldots, k$ and

$D_{k+1}, \ldots, D_N$ do not intersect U. (This is the definition of simple normal crossings.) Then in the inequality we have just proved for ψ, we may sum just for $j = 1, \ldots, k$ to obtain

$$(*) \qquad \psi \geq \omega + \sum_{j=1}^{k} \frac{\sqrt{-1}}{\pi} \frac{\partial \log|\alpha s_j|^2 \wedge \bar{\partial} \log|\alpha s_j|^2}{(\log|\alpha s_j|^2)^2} > 0.$$

Each term in the sum on the right is a (1,1)-form ≥ 0. For each $j = 1, \ldots, k$ there is a C^∞ function $\gamma_j(z) > 0$ such that

$$|s_j|^2 = \gamma_j(z)|z_j|^2.$$

Directly from the definition of ∂ and $\bar{\partial}$ we find:

$$\partial \log|\alpha s_j|^2 \wedge \bar{\partial} \log|\alpha s_j|^2 = \frac{dz_j \wedge d\bar{z}_j}{|z_j|^2} + \varphi_j,$$

where

$$\varphi_j = \frac{\partial \gamma_j \wedge \bar{\partial} \gamma_j}{\gamma_j^2} + \frac{\partial \gamma_j \wedge d\bar{z}_j}{\bar{z}_j \gamma_j} + \frac{dz_j \wedge \bar{\partial} \gamma_j}{z_j \gamma_j}.$$

We note that φ_j is a sum of three terms. The first behaves in a different way from the second and third, while the second and third behave essentially the same way from whatever point of view we shall consider.

At the cost of shrinking U if necessary, we can find a constant $c > 0$ such that

$$\omega \geq c \sum_{j=1}^{n} \frac{\sqrt{-1}}{2\pi} dz_j \wedge d\bar{z}_j.$$

Let $g_j = |z_j|^2 (\log|\alpha s_j|^2)^2$. Then we obtain

$$(**) \qquad \psi \geq c \sum_{j=1}^{n} \frac{\sqrt{-1}}{2\pi} dz_j \wedge d\bar{z}_j + \sum_{j=1}^{k} \frac{\sqrt{-1}}{\pi} \frac{dz_j \wedge d\bar{z}_j}{g_j}$$

$$+ \sum_{j=1}^{k} \frac{\sqrt{-1}}{\pi} \frac{\partial \gamma_j \wedge \bar{\partial} \gamma_j}{\gamma_j^2 (\log|\alpha s_j|^2)^2}$$

$$+ \sum_{j=1}^{k} \frac{\sqrt{-1}}{\pi} \left[z_j \frac{\partial \gamma_j \wedge d\bar{z}_j}{\gamma_j g_j} + \bar{z}_j \frac{dz_j \wedge \bar{\partial} \gamma_j}{\gamma_j g_j} \right].$$

We note that on the right-hand side of the inequality, we merely rewrote the expression of a positive form, but the individual terms in the last sum need not be positive. We now use Lemma 2.4, and get

$$\psi^n \geq (\text{right-hand side})^n$$

because the right-hand side is non-negative. We shall prove that

$$(\text{right-hand side})^n \geq c_1 \frac{\Phi}{g} + \frac{\Lambda}{g},$$

where Φ is the euclidean form, and $\Lambda(z)$ is C^∞ and vanishes on the divisor D. The n-th power of the right-hand side will be a sum of cross terms with positive coefficients, and these terms have the form (up to a constant factor)

$$\prod_{j\in J_1} \sqrt{-1}\, dz_j \wedge d\bar{z}_j \prod_{j\in J_2} \sqrt{-1}\, \frac{dz_j \wedge d\bar{z}_j}{g_j}$$

$$\prod_{j\in J_3} \sqrt{-1}\, \frac{\partial\gamma_j \wedge \bar{\partial}\gamma_j}{\gamma_j^2(\log|\alpha s_j|^2)^2} \prod_{j\in J_4} \sqrt{-1} \left[z_j \frac{\partial\gamma_j \wedge d\bar{z}_j}{\gamma_j g_j} + \bar{z}_j \frac{dz_j \wedge \bar{\partial}\gamma_j}{\gamma_j g_j} \right].$$

The sets of indices J_1, J_2, J_4 are mutually disjoint, and J_3, J_4 are mutually disjoint, because of the alternating product. However, J_3 can have elements in common with J_1 or J_2. The sets J_2, J_3 and J_4 are subsets of $\{1,\ldots,k\}$. There is one term when J_3 and J_4 are empty, $J_2 = \{1,\ldots,k\}$ and $J_1 = \{k+1,\ldots,n\}$. This is the main term, and since g_j is C^∞ nonvanishing at the given point for $j = k+1,\ldots,n$ it follows that this main term satisfies the desired inequality

$$c \prod_{j=k+1}^{n} \sqrt{-1}\, dz_j \wedge d\bar{z}_j \prod_{j=1}^{k} \sqrt{-1}\, \frac{dz_j \wedge d\bar{z}_j}{g_j} \geq B\Psi_\alpha$$

on some neighborhood of our given point.

If a term occurs without any index $j \in J_4$ then this term is a non-negative form which improves our inequality. Suppose there is a term with some $j \in J_4$. Then this term can be written as

$$\frac{\Phi(z)}{g} h(z) \quad \text{with} \quad g = \prod_{j=1}^{N} |s_j|^2(\log|\alpha s_j|^2)^2$$

and where h is continuous and vanishes on $D \cap U$; this is due to $(\log|\alpha s_j|^2)^2 \to \infty$ on D_j, and z_j or $\bar{z}_j = 0$ on D_j. If we shrink U further, this implies that all the other terms can be made small relative to the main term, and therefore that

$$\psi^n \geq (B - \varepsilon)\Psi_\alpha$$

on some neighborhood of the given point for all sufficiently small α. This concludes the proof of the theorem.

Examples. Suppose first that $X = \mathbf{P}^1$ and D consists of three distinct points. Then the above construction shows that the complement of the three points admits a strongly hyperbolic (1,1)-form, and therefore is hyperbolic. This is one way to bypass the uniformization theorem, say to prove Picard's theorem.

Similarly, let X be a compact Riemann surface of genus p, and let D consists of $N > 2 - 2p$ points. Then the complement of D admits a strongly hyperbolic (1,1)-form.

For other constructions of positive (1,1)-forms on surfaces, see Grauert-Reckziegel [G–R].

Finally, let D consist of $n + 2$ hyperplanes in general position in $\mathbf{P}^n$. We recall that general position means that any $n + 1$ are linearly independent. Then this implies that D has simple normal crossings. Furthermore, the canonical class is that of $-(n + 1)H$, where H is a hyperplane, so $D - (n + 1)H$ is ample, and therefore satisfies the conditions of the theorem. Again, we conclude that the complement of $n + 2$ hyperplanes in general position in $\mathbf{P}^n$ admits a strongly hyperbolic volume form. This complement is therefore measure hyperbolic. The extent to which it is not hyperbolic is discussed in Chapter VII.

Although we prove the existence of a hyperbolic volume form above under some assumption, there remains the Kobayashi problem in this context (equidimensional): If X is a projective non-singular variety and is measure hyperbolic, does there exist such a form? In the next section, we discuss this context further.

IV, §5. PSEUDO CANONICAL VARIETIES

This section deals with an application to algebraic geometry and uses some (elementary) facts about cohomology of sheaves on such varieties, but nothing beyond the basics of Hartshorne's book, Chapter III. Readers can omit this section without impairing the understanding of the rest of the book.

Also the theorems of this section stem from Griffiths [Gri 2], Kobayashi-Ochiai [K–O 4], with an appendix by Kodaira which shows how results proved for an ample canonical class extend to the pseudo ample case (to be defined shortly).

Let L be a line bundle on a compact complex manifold X. We abbreviate the tensor products $L^{\otimes m}$ by L^m. We say that L is **ample** if there exists some positive integer m such that a basis of sections $(s_0, \ldots, s_N)$ of $H^0(X, L^m)$ generates L^m at every point, and give a projective imbedding

$$\varphi_m \colon (s_0, \ldots, s_N) \colon X \to \mathbf{P}^N.$$

We say that L is **very ample** if we can take $m = 1$ in the above condition, so already sections of $H^0(X, L)$ give the imbedding. If the above condition holds for some $m = m_0$, then it holds for all m multiples of m_0.

Even if we do not get a projective imbedding by means of a basis for the sections, we still get a rational map into $\mathbf{P}^N$. We let

$$m \underset{\text{div}}{\to} \infty$$

denote the property that m tends to infinity ordered by divisibility. We say that L is **pseudo ample** if φ_m is birational for some positive m_0, and thus gives a projective imbedding of a non-empty Zariski open subset of X, for m large ordered by divisibility.

Let K_X be the canonical bundle,

$$K_X = \bigwedge\nolimits^{\text{top}} T^{\vee} X.$$

Classically, X has been called canonical if K_X is very ample [Gri 2]. But for the same reason that Grothendieck changed the meaning of "ample" to what it is now, it seems more fruitful to say that X is **canonical** if K_X is ample, and **very canonical** if K_X is very ample. On the other hand, if K_X is pseudo ample, then X is usually said to be of **general type**, but with the support of Griffiths, I shall say that X is **pseudo cannonical** to make the terminology functorial with respect to the ideas.

We now suppose that X is algebraic, that it has a projective imbedding in projective space. By a **variety** we shall mean an irreducible algebraic set in projective space over $\mathbf{C}$. We recall an elementary result:

Let X, Y be non-singular varieties. Then a birational map $X \to Y$ induces an isomorphism

$$H^0(X, K_X^m) \to H^0(Y, K_Y^m)$$

for every positive integer m.

Proof. See Hartshorne, Chapter II, Theorem 8.19. The proof given there when $m = 1$ works in general. The idea is that a section of K_X^m can be represented globally birationally, for instance as

$$g(dg_1 \wedge \cdots \wedge dg_n)^m,$$

where $g, g_1, \ldots, g_n$ are rational functions. If $g_1, \ldots, g_n$ are local parameters at a point, then g is in the local ring at that point. One now uses the fact that if a rational function is not in the local ring of a point, then it has a divisorial pole passing through that point. Such a pole induces a point on any complete model X or Y. This shows that if a rational form as above gives rise to a section of K_X^m for one model, then it must give rise to a section of K_Y^m for any other model.

We shall be interested in the dimensions

$$h^0(X, L^m) = \dim H^0(X, L^m)$$

for various line bundles L, starting with the canonical bundle, but involving other bundles as well. In speaking of estimates, we use the standard notation of number theorists

$$A(m) \ll B(m) \qquad \text{for} \quad m \to \infty$$

to mean that there is a constant c such that $A(m) \leq cB(m)$ for all m sufficiently large. If the going to infinity is by divisibility, then sufficiently large is according to this ordering. Following Kodaira's addendum to [K–O 4], we shall now construct a pseudo volume form on a pseudo canonical projective complex manifold, with positive Ricci form. We recall two lemmas from basic algebraic geometry.

Lemma 5.1. *Let X be a projective variety of dimension n. Let D be a divisor on X. Let $h^0(mD) = h^0(\mathcal{O}_X(mD))$. Then*

$$h^0(mD) = \dim H^0(X, mD) \ll m^n \qquad \text{for} \quad m \to \infty.$$

Proof. Let E be a divisor which is ample and such that $D + E$ is ample. Then we have an inclusion

$$H^0(mD) \subset H^0(mD + mE), \qquad \text{so} \qquad h^0(mD) \leq h^0(mD + mE).$$

Furthermore, if $E' = D + E$ is ample, then

$$h^0(mE') = \chi(mE') \qquad \text{for } m \text{ large,}$$

because the higher cohomology groups vanish for m large, and the Euler characteristic is a polynomial in m of degree $\leq n$, thus proving the lemma.

Lemma 5.2. *Let X be a projective non-singular variety of dimension n. Let E be very ample on X, and let D be a divisor on X such that*

$$h^0(mD) \gg m^n \qquad \text{for} \quad m \underset{\text{div}}{\to} \infty.$$

Then

$$h^0(mD - E) \gg m^n \quad for \quad m \xrightarrow[\text{div}]{} \infty.$$

Proof. Without loss of generality, we may replace E by any divisor in its class, and thus we may assume that E is an irreducible non-singular subvariety of X (a hyperplane section, in fact). We have the exact sequence

$$0 \to \mathcal{O}(mD - E) \to \mathcal{O}(mD) \to \mathcal{O}(mD)|E \to 0$$

whence the exact cohomology sequence

$$0 \to H^0(X, mD - E) \to H^0(X, mD) \to H^0\big(E, (\mathcal{O}(D)|E)^m\big).$$

Applying the first lemma to the invertible sheaf $\mathcal{O}(mD)|E$ on E, we conclude that the dimension of the term on the right is $\ll m^{n-1}$, so $h^0(X, mD - E) \gg m^n$ for m large, and in particular is positive for m large, whence the lemma follows.

Suppose that X is pseudo canonical. Then $h^0(X, K_X^m) \gg m^n$ for m large (ordered by divisibility), so we can apply the lemma. Let L be a very ample line bundle on X. We shall obtain a projective imbedding of X by means of *some* of the sections in $H^0(X, K_X^m)$. By Lemma 5.2, for m large there exists a non-trivial holomorphic section α of $K_X^m \otimes L^{-1}$. Let $\{s_0, \ldots, s_N\}$ be a basis of $H^0(X, L)$. Then

$$\alpha \otimes s_0, \ldots, \alpha \otimes s_N$$

are linearly independent sections of $H^0(X, K_X^m)$. Since $(s_0, \ldots, s_N)$ gives a projective imbedding of X into $\mathbf{P}^N$ because L is assumed very ample, it follows that $\alpha \otimes s_0, \ldots, \alpha \otimes s_N$ vanish simultaneously only at the zeros of α, but nevertheless give the *same projective imbedding*, which is determined only by their ratios. Then

$$\alpha \bar{\alpha} \sum s_j \otimes \bar{s}_j$$

may be considered as a section of

$$(K_X^m L^{-1})L \otimes (\bar{K}_X^m \bar{L}^{-1})\bar{L} = K_X^m \otimes \bar{K}_X^m,$$

and can be locally expressed in terms of complex coordinates in the form

$$|g(z)|^2 \sum_{j=1}^{n} |g_j(z)|^2 \Phi(z)^{\otimes m},$$

where as usual $\Phi(z)$ is the euclidean volume form on $\mathbf{C}^n$, while $g(z)$, $g_0(z),\ldots,g_N(z)$ are local holomorphic functions representing α, $s_0,\ldots,s_N$ respectively. The following is then immediate:

Let

$$h(z) = \left(\sum_{j=1}^{n} |g_j(z)|^2 \right)^{1/m},$$

Then there is a unique pseudo volume form Ψ on X which has the local expression

$$\Psi(z) = |g(z)|^{2/m} h(z) \Phi(z).$$

Furthermore $\mathrm{Ric}(\Psi)$ is positive, because $\mathrm{Ric}(\Psi)$ is the pullback of the Fubini-study form on $\mathbf{P}^N$ by the projective imbedding.

In particular, we have proved:

Theorem 5.3 (Kodaira, Kobayashi–Ochiai). *Let X be a non-singular pseudo canonical variety. Then X admits a pseudo volume form Ψ with $\mathrm{Ric}(\Psi)$ positive, and X is measure hyperbolic.*

The converse statements are conjectural.

Conjecture 5.4 (Kobayashi [Ko 4], Chapter IX). *If X is measure hyperbolic, then X is pseudo canonical.*

This would give the neat statement that:

A non-singular variety is pseudo canonical if and only if it is measure hyperbolic.

Kobayashi's conjecture is known for surfaces, through the paper of Green–Griffiths [G–G], completed in one remaining case (arising from the classification of surfaces) by Bogomolov–Mumford. Cf. the appendix of Mori–Mukai [M–M].

On the other hand, the converse in the differential geometric context is also a problem raised by Kobayashi [Ko 6], Theorem 7.1 and p. 377), namely:

If X is non-singular, and there exists a pseudo volume form with positive Ricci form, is X pseudo canonical?

[*Added in proof.* Burt Totaro has proved this conjecture recently. In fact he has proved that if a line bundle admits a pseudo metric with positive Chern form then it is pseudo ample.]

I would conjecture:

Conjecture 5.5. *A variety X is hyperbolic if and only if every subvariety (including X itself) is pseudo canonical.*

Comparison between the equidimensional case and the case of holomorphic maps $f: \mathbf{C} \to X$ constitutes one of the main aspects of the theory of hyperbolic spaces. See [La 3].

Finally I make some brief comments on the direct relation between the property of being pseudo canonical and Brody hyperbolicity.

In [La 3] I defined the **exceptional set** $\operatorname{Exc}(X)$ of a projective variety X to be the Zariski closure of the union of all images of non-constant holomorphic maps $\mathbf{C} \to X$. This exceptional set may or may not be the whole variety, and the reader will find a discussion of conjectures attempting to characterize those varieties for which the exceptional set is a proper subset, and attempting to describe its structure. We have the following possibilities:

Conjecture 5.6. *The exceptional set is a proper subset if and only if X is pseudo canonical.*

Conjecture 5.7. *The exceptional set is the Zariski closure of the union of all non-constant rational images of $\mathbf{P}^1$ and abelian varieties into X.*

This last property would give an algebraic characterization of the exceptional set, and would show that the most algebraic and essentially weakest way of defining it coincides with the strongest way of defining it. It would also show that the only holomorphic maps of $\mathbf{C}$ into X arise from the "obvious" structure of X allowing such maps. Conjecture 5.7 was stated in [La 3] without taking the Zariski closure, but I now believe it is necessary to take this closure, especially in higher dimensions.

One can also pseudofy the notion of hyperbolicity. We say that X is **pseudo Kobayashi hyperbolic** if there exists a proper algebraic subset Y such that if $x, x' \in X$ and $d_X(x, x') = 0$ then $x = x'$ or $x, x' \in Y$. We say that X is **pseudo Brody hyperbolic** if there exists a proper algebraic subset Z such that every non-constant holomorphic map of $\mathbf{C}$ into X actually maps $\mathbf{C}$ into Z.

Conjecture 5.8. *A projective variety X is pseudo Kobayashi hyperbolic if and only if X is pseudo Brody hyperbolic. The sets Y and Z above can be taken to be the exceptional set $\operatorname{Exc}(X)$.*

In light of the second statement in this conjecture, we could also say that X is hyperbolic modulo the exceptional set, cf. Chapter II, §1.

CHAPTER V

Curvature on Vector Bundles

In this chapter, we extend some of the one-dimensional notions of Chern and Ricci forms to vector bundles. First we do this in the hermitian case. The basic reference is Griffiths' positivity paper [Gri 1], which cleared up a lot of the formalism in this case. We shall give also another interpretation of the Griffiths function on imbedded complex submanifolds of dimension 1, coming from the higher dimensional tangent bundle, due to Wu.

Kobayashi [Ko 5] showed how the differential geometric formalism of connections extends to vector bundles with a length function which is smooth outside the zero section. However, Garrity [Ga] gave a very neat differential geometric characterization for "negative" vector bundles, and we shall reproduce his result in the last section, which is logically independent of the rest of the chapter. For many purposes, I recommend reading it first.

V, §1. CONNECTIONS ON VECTOR BUNDLES

This section consists principally of linear algebra, and is basically general (anti) commutative algebra. But we place ourselves immediately in the complex vector bundle case to fit our applications.

Let X be a complex manifold and let $E \to X$ be a complex vector bundle. For each open set U of X we let E_U be the restriction of E to U. We suppose E has rank r, by which we mean that its fibers have dimension r.

By a **frame** $e = (e_1, \ldots, e_r)$ we mean r sections over an open set U such that $e_1(x), \ldots, e_r(x)$ is a basis of the fiber E_x at each point $x \in U$. **Sections** are assumed C^∞ unless otherwise specified. We use the following notation:

$\mathcal{A}^p(X) = $ sheaf of p-forms.

$\mathcal{A}^0(E) = $ sheaf of sections of E.

$\mathcal{A}^p(E) = \mathcal{A}^p(X) \otimes \mathcal{A}^0(E)$ (tensor product over $\mathcal{A}^0(X)$).

Thus an element of $\mathcal{A}^p(E)(U)$ can be expressed as a sum

$$\omega = \sum_{j=1}^{N} \omega_j \otimes s_j,$$

where $\omega_j \in \mathcal{A}^p(X)(U)$ is a differential form over U, and $s_j \in \mathcal{A}^0(E)(U)$ is a section over U. If e is a frame over U, then ω can be expressed *uniquely*

$$\omega = \sum_{i=1}^{r} \omega_i \otimes e_i.$$

If we let

$$\mathcal{A}^\cdot(X) = \bigoplus \mathcal{A}^p(X) \qquad \text{and} \qquad \mathcal{A}^\cdot(E) = \bigoplus \mathcal{A}^p(E),$$

then $\mathcal{A}^\cdot(E)$ is a graded sheaf module over the graded sheaf algebra $\mathcal{A}^\cdot(X)$, under the natural wedge product on the left.

A **connection** on E is a **C**-linear map

$$D^0 \colon \mathcal{A}^0(E) \to \mathcal{A}^1(E)$$

satisfying

$$D^0(fs) = df \otimes s + f D^0 s \qquad \text{for all} \qquad f \in \mathcal{A}^0(X)(U), \quad s \in \mathcal{A}^0(E)(U).$$

We suppose given a connection.

Proposition 1.1. *For each $p \geq 1$ there exists a unique* **C**-*linear map*

$$D^p \colon \mathcal{A}^p(E) \to \mathcal{A}^{p+1}(E)$$

such that

$$D^p(\omega \otimes s) = d\omega \otimes s + (-1)^p \omega \wedge Ds$$

for all $\omega \in \mathcal{A}^p(X)(U)$ and $s \in \mathcal{A}^0(E)(U)$.

Proof. Immediate from the universality of the tensor product.

We shall often omit the superscripts and write D for the map on the graded sheaf

$$D: \bigoplus \mathscr{A}^p(E) \to \bigoplus \mathscr{A}^p(E).$$

Proposition 1.2. *For* $\omega \in \mathscr{A}^p(X)(U)$ *and* $\psi \in \mathscr{A}^q(E)(U)$ *we have:*

$$D^{p+q}(\omega \wedge \psi) = d\omega \wedge \psi + (-1)^p \omega \wedge D^q \psi,$$

Proof. To carry out the proof, let rather $\psi \in \mathscr{A}^q(X)(U)$ and $s \in \mathscr{A}^0(E)(U)$. Then

$$D(\omega \wedge \psi \otimes s) = d(\omega \wedge \psi) \otimes s + (-1)^{p+q} \omega \wedge \psi \wedge Ds$$

$$= d\omega \wedge \psi \otimes s + (-1)^p \omega \wedge d\psi \otimes s + (-1)^{p+q} \omega \wedge \psi \wedge Ds$$

$$= d\omega \wedge \psi \otimes s + (-1)^p \omega \wedge [d\psi \otimes s + (-1)^q \psi \wedge Ds]$$

which proves the proposition.

Proposition 1.3. *For all* $p, q \geq 0$ *the composite* $D^{p+1} \circ D^p$ *is* $\mathscr{A}^p(X)$-*linear, that is for* $\omega \in \mathscr{A}^p(X)(U)$ *and* $\psi \in \mathscr{A}^q(E)(U)$,

$$D \circ D(\omega \wedge \psi) = \omega \wedge D \circ D\psi.$$

Proof. We have using Proposition 1.2:

$$D^{p+q+1} D^{p+q}(\omega \wedge \psi) = D^{p+q+1}[d\omega \wedge \psi) + (-1)^p \omega \wedge D^q \psi]$$

$$= dd\omega \wedge \psi + (-1)^{p+1} d\omega \wedge D\psi + (-1)^p d\omega \wedge D\psi$$

$$+ (-1)^p(-1)^p \omega \wedge D^{q+1} D^q \psi$$

$$= \omega \wedge D^{q+1} D^q \psi$$

as was to be shown.

Next we tabulate the matrix relations with respect to a frame. For any section

$$s = \sum f_i e_i \qquad \text{with} \quad f_i \in \mathscr{A}^0(X)(U),$$

in other words, with functions $f_1, \ldots, f_r$ we write the coordinate vector

$$s_e = \begin{pmatrix} f_1 \\ f_2 \\ \vdots \\ f_r \end{pmatrix} \qquad \text{and} \qquad e = (e_1, \ldots, e_r).$$

Thus

$$s = es_e = {}^t s_e \, {}^t e.$$

Similarly for differential forms,

$$\omega = {}^t \omega_e \otimes {}^t e = \sum_{i=1}^{r} \omega_i \otimes e_i, \qquad \text{where} \quad \omega_e = \begin{pmatrix} \omega_1 \\ \omega_2 \\ \vdots \\ \omega_r \end{pmatrix}.$$

We note that $\{e_1, \ldots, e_r\}$ is a basis of $\mathcal{A}^p(E)(U)$ over $\mathcal{A}^p(X)(U)$, and the column vector ω_e is called the vector of **coordinates** of ω with respect to the frame. It is a vertical vector whose components are differential forms in $\mathcal{A}^p(X)(U)$. Sometimes, by abuse of notation, we also write

$$\omega = e \omega_e$$

on the "wrong" side, to avoid writing transpose signs.

The connection applied to a section can be written

$$Ds = \sum df_i \otimes e_i + \sum_{i,j} f_j \theta_{ij} \otimes e_i,$$

where $(\theta_{ij}) = \theta_e$ is a matrix of 1-forms. Then for $\psi \in \mathcal{A}^p(E)(U)$,

$$(D\psi)_e = (d + \theta_e)\psi_e = \begin{pmatrix} d\psi_1 \\ \vdots \\ d\psi_r \end{pmatrix} + \begin{pmatrix} \theta_{11} & \cdots & \theta_{1r} \\ \vdots & & \vdots \\ \theta_{r1} & \cdots & \theta_{rr} \end{pmatrix} \begin{pmatrix} \psi_1 \\ \vdots \\ \psi_r \end{pmatrix}.$$

Thus we may write symbolically

1.4. $D_e = d + \theta_e.$

This formula holds in all degrees.

The composite $D \circ D$ is called the **curvature operator**, or simply **curvature, of the connection**, for god knows what reasons, it's an impediment to understand the simple linear algebra involved. We also write

$$\Theta = DD \qquad \text{and} \qquad \Theta_e = \text{matrix of } \Theta \text{ with respect to a frame.}$$

Thus Θ_e is a matrix of 2-forms, given by $(d + \theta_e)^2$. We claim:

1.5. $\Theta_e = d\theta_e + \theta_e \wedge \theta_e.$

Note that unless $r = 1$, the wedge product $\theta_e \wedge \theta_e$ is not zero, since it involves multiplication of matrices, so single coordinate forms are repeated. The proof of **1.5** is as easy as previous proofs, namely:

$$
\begin{aligned}
(d + \theta_e)^2 \psi_e &= d(d\psi_e + \theta_e \wedge \psi_e) + \theta_e \wedge (d\psi_e + \theta_e \wedge \psi_e) \\
&= d(\theta_e \wedge \psi_e) + \theta_e \wedge d\psi_e + \theta_e \wedge \theta_e \wedge \psi_e \\
&= d\theta_e \wedge \psi_e - \theta_e \wedge d\psi_e + \theta_e \wedge d\psi_e + \theta_e \wedge \theta_e \wedge \psi_e \\
&= d\theta_e \wedge \psi_e + \theta_e \wedge \theta_e \wedge \psi_e
\end{aligned}
$$

as was to be shown.

In any graded algebra $A = \bigoplus A^p$ one has the **graded bracket symbol** (**graded commutator**)

$$
[a, b] = ab - (-1)^{pq} ba \qquad \text{for} \quad a \in A^p \quad \text{and} \quad b \in A^q.
$$

Unless otherwise specified, this is the symbol we use. Then we have

1.6. $d\Theta_e = [\Theta_e, \theta_e]$.

Proof. On the one hand, from **1.5** we get (omitting e for simplicity):

$$
d\Theta = d\theta \wedge \theta - \theta \wedge d\theta.
$$

On the other hand,

$$
\begin{aligned}
[\Theta, \theta] &= [d\theta + \theta \wedge \theta, \theta] \\
&= d\theta \wedge \theta + \theta \wedge \theta \wedge \theta - (-1)^2 (\theta \wedge d\theta + \theta \wedge \theta \wedge \theta) \\
&= d\theta \wedge \theta - \theta \wedge d\theta \\
&= d\Theta
\end{aligned}
$$

as was to be shown. [In this case, the graded bracket was the usual bracket since $pq = 2$.]

Next we give the formula showing how the various objects change under a change of frame. Since for $g \in \mathrm{GL}_r(\mathscr{A}^0(X)(U))$ one can write

$$
e\omega_e = egg^{-1}\omega_e,
$$

we have the relation

1.7. $\omega_{eg} = g^{-1}\omega_e$.

Then for the connection matrix, we obtain

1.8. $\theta_{eg} = g^{-1}dg + g^{-1}\theta_e g.$

This comes from writing

$$Ds = egg^{-1}d(gg^{-1}s_e) + egg^{-1}\theta_e gg^{-1}s_e,$$

as in elementary linear algebra, and $d(gg^{-1}s_e) = (dg)g^{-1}s_e + gd(g^{-1}s_e)$. Finally for the curvature we get:

1.9. $\Theta_{eg} = g^{-1}\Theta_e g.$

This is immediate, like the previous relations.

As an application of relation **1.8**, we obtain the often used

Lemma 1.10. *Given a connection D and a point x, there exists a frame e such that $\theta_e(x) = 0$.*

Proof. Let $z = (z_1, \ldots, z_n)$ be holomorphic coordinates. Write

$$g(z) = I + M(z) \quad \text{so} \quad g(0) = I, \quad M(0) = O.$$

By **1.8** we know that

$$\theta_{eg}(0) = \theta_e(0) + g^{-1}(0)(dg)(0).$$

We want to solve this equation in $g(z)$ to make $\theta_{eg}(0) = 0$. Thus we have to solve for $M(z)$ such that

$$dM(0) = -\theta_e(0)$$

which can be done trivially, linearly, by integration because it is a local problem. This proves the lemma. (But see Lemma 2.6.)

Tensor products

Let E, F be two vector bundles over X. Let D_E, D_F be connections on E and F respectively. Then there exists a unique connection D on $E \otimes F$ such that for all sections $s \in \mathcal{A}^0(E)(U)$ and $t \in \mathcal{A}^0(F)(U)$ we have

$$D(s \otimes t) = D_E s \otimes t + s \otimes D_F t.$$

This is immediate from the definitions.

Dual bundle

*Let $E^{\vee}$ be the (complex) dual bundle of E. Let D be a connection on E.
Then there exists a unique connection $D^{\vee}$ on $E^{\vee}$ such that for
$s \in \mathscr{A}^0(E)(U)$ and $t \in \mathscr{A}^0(E^{\vee})(U)$ we have*

$$d\langle t, s \rangle = \langle D^{\vee}t, s \rangle + \langle t, Ds \rangle.$$

Again this is immediate from punctual duality. Note that there is a
natural pairing

$$\mathscr{A}^p(E^{\vee}) \times \mathscr{A}^q(E) \to \mathscr{A}^{p+q}(X)$$

such that for $\omega \in \mathscr{A}^p(X)(U)$, $\psi \in \mathscr{A}^q(X)(U)$, $s \in \mathscr{A}^0(E)(U)$, $t^{\vee} \in \mathscr{A}^0(E^{\vee})(U)$
we have

$$(\omega \otimes t^{\vee}) \times (\psi \otimes s) \mapsto \omega \wedge \psi \langle t^{\vee}, s \rangle,$$

where $\langle t^{\vee}, s \rangle$ is the evaluation of the section $t^{\vee}$ in the dual on the section s of E. Thus $\langle t^{\vee}, s \rangle$ is a function on U.

We can combine the above two constructions and consider the vector
bundle

$$\operatorname{Hom}(E, E) \approx E^{\vee} \otimes E,$$

which has fiber dimension r^2 if E has rank r. Let

$$\Gamma \in \mathscr{A}^p(\operatorname{Hom}(E, E))(U).$$

Then a frame e establishes an isomorphism

$$\mathscr{A}^p(E)(U) \xrightarrow{\approx} \mathscr{A}^p(X)(U)^{(r)}$$

and Γ is represented by an $r \times r$ matrix of p-forms Γ_e, making the following diagram commutative.

$$
\begin{array}{ccc}
\mathscr{A}^0(E)(U) & \xrightarrow{\ \Gamma\ } & \mathscr{A}^p(E)(U) \\
\approx \downarrow & & \downarrow \approx \\
\mathscr{A}^0(X)(U)^{(r)} & \xrightarrow[\Gamma_e]{} & \mathscr{A}^p(X)(U)^{(r)}
\end{array}
$$

From first principles, we have

1.11. $\Gamma_{eg} = g^{-1}\Gamma_e g.$

We now see that the curvature Θ is precisely such an element of $H^0(\mathscr{A}^2(\operatorname{Hom}(E, E)))$, satisfying this transformation law.

Finally we look at connection and curvature matrices on the dual bundle and $\operatorname{Hom}(E, E)$. Let e be a frame and $e^\vee$ the dual frame. By this we mean that if $e = (e_1, \ldots, e_r)$ and $e^\vee = (e_1^\vee, \ldots, e_r^\vee)$ then

$$\langle e_i^\vee, e_j \rangle = \delta_{ij}.$$

It follows that

$$0 = \langle D^\vee e_i^\vee, e_j \rangle + \langle e_i, De_j \rangle.$$

By definition,

$$D^\vee e_i^\vee = \sum_k \theta_{ki}^\vee \otimes e_k^\vee \qquad \text{and} \qquad De_j = \sum_k \theta_{kj} \otimes e_k.$$

Therefore if $\theta_e^\vee$ is the matrix of $D^\vee$ with respect to $e^\vee$, we get

1.12. $\theta_{ij}^\vee = -\theta_{ji}$, or $\theta_e^\vee = -{}^t\theta_e$.

Next let $\eta \in \mathscr{A}^p(\operatorname{Hom}(E, E))(U) \approx \mathscr{A}^p(E^\vee \otimes E)(U)$. We can write

$$\eta = \sum \eta_{ji} e_i^\vee \otimes e_j \qquad \text{with} \quad \eta_{ji} \in \mathscr{A}^p(X)(U).$$

We let

$$\eta_e = \operatorname{matrix}(\eta_{ji}),$$

omitting the $e^\vee$ from this notation to simplify indices. Then

$$D\eta = d\eta + (-1)^p \sum \eta_{ji} \wedge D^\vee e_i^\vee \otimes e_j + (-1)^p \sum \eta_{ji} e_i^\vee \wedge De_j.$$

Using **1.12** and ${}^t({}^t\eta \wedge {}^t\theta) = (-1)^p \theta \wedge \eta$, we obtain the formula

1.13. $D\eta_e = d\eta_e + \theta \wedge \eta_e - (-1)^p \eta_e \wedge \theta = d\eta_e + [\theta, \eta_e]$.

Proposition 1.14 (Bianchi identity). *Let Θ be the curvature of a connection D. Then*

$$D\Theta = 0.$$

Proof. We apply **1.13** with $\eta = \Theta$. By **1.6** we know that

$$d\Theta_e = \Theta_e \wedge \theta_e - \theta_e \wedge \Theta_e$$

and since $p = 2$ this is just what is needed to cancel the other two terms in **1.13**, as was to be shown.

V, §2. COMPLEX HERMITIAN CONNECTIONS AND RICCI TENSOR

A vector bundle is said to be **hermitian** if it is endowed with a hermitian structure, i.e. a positive definite hermitian form at each point varying smoothly over the manifold.

A vector bundle is said to be **holomorphic** if it is given by an atlas with holomorphic charts, i.e. the chart transformations given locally by a map

$$g: U \to \mathrm{GL}_r(\mathbf{C})$$

are holomorphic.

Warning. In Griffiths–Harris, a hermitian bundle is assumed automatically to be holomorphic. This is not a universal convention, and I won't adopt it, at the cost of using an extra word each time. I can hardly resist coining the word holomitian to combine the two, however.

Let X be a complex manifold and E a complex vector bundle over X. To begin with, we have a direct sum decomposition

$$\mathscr{A}^m(X) = \bigoplus_{p+q=m} \mathscr{A}^{p,q}(X),$$

where $\mathscr{A}^{p,q}(X)(U)$ consists of those differential forms which can be expressed as a sum of terms

$$\sum f_{IJ}(z, \bar{z})\, dz_{i_1} \wedge \cdots \wedge dz_{i_p} \wedge d\bar{z}_{j_1} \wedge \cdots \wedge d\bar{z}_{j_q}$$

and $i_1 < \cdots < i_p$, $j_1 < \cdots < j_q$. A holomorphic change of charts on X shows that the pair (p, q) does not depend on the chart.

We then define

$$\mathscr{A}^{p,q}(E) = \mathscr{A}^{p,q}(X) \otimes \mathscr{A}^0(E).$$

We have the operators

$$\partial: \mathscr{A}^{p,q}(X) \to \mathscr{A}^{p+1,q}(X) \qquad \text{and} \qquad \bar{\partial}: \mathscr{A}^{p,q}(X) \to \mathscr{A}^{p,q+1}(X),$$

where for instance locally

$$\bar{\partial}(f(z, \bar{z})\, dz_I \wedge d\bar{z}_J) = \sum_{k=1}^{n} \frac{\partial f}{\partial \bar{z}_k}\, d\bar{z}_k \wedge dz_I \wedge d\bar{z}_J.$$

If we now assume that E is a holomorphic vector bundle, then we can define the operator $\bar{\partial}$ on $\mathscr{A}^{p,q}(E)$ as follows. Let e be a holomorphic frame, and let $\omega \in \mathscr{A}^{p,q}(E)(U)$. Let ω_e as before be its coordinate vector with respect to e, with components in $\mathscr{A}^{p,q}(X)(U)$. There is a unique form $\bar{\partial}\omega$ such that

$$(\bar{\partial}\omega)_e = \bar{\partial}\omega_e.$$

Indeed, if $e \mapsto eg$ is a change of holomorphic frames, with g holomorphic, then $\omega_{eg} = g^{-1}\omega_e$ and

$$\bar{\partial}(g^{-1}\omega_e) = g^{-1}\bar{\partial}\omega_e$$

because $\bar{\partial}g^{-1} = 0$ by Cauchy–Riemann. Thus we have the map

$$\bar{\partial}: \mathscr{A}^{p,q}(E) \to \mathscr{A}^{p,q+1}(E)$$

which is well defined.

Let E be a holomorphic vector bundle. Then a connection

$$D: \mathscr{A}^0(E) \to \mathscr{A}^1(E)$$

can be decomposed into a direct sum because

$$\mathscr{A}^1(E) = \mathscr{A}^{1,0}(E) \oplus \mathscr{A}^{0,1}(E).$$

We let

$$D': \mathscr{A}^0(E) \to \mathscr{A}^{1,0}(E) \qquad \text{and} \qquad D'': \mathscr{A}^0(E) \to \mathscr{A}^{0,1}(E)$$

be its two components. The connection is said to be **compatible with the complex structure**, or a **complex connection**, if

$$D'' = \bar{\partial}$$

In this case, we can write

$$D = D' + \bar{\partial}.$$

Suppose in addition that E is hermitian, with hermitian product $\langle \, , \, \rangle$ antilinear in its second variable. We say that D is **compatible with the hermitian structure**, or a **hermitian connection**, if

$$d\langle s, t \rangle = \langle Ds, t \rangle + \langle s, Dt \rangle$$

for all $s, t \in \mathscr{A}^0(E)(U)$.

Theorem 2.1. *Let* E *be a hermitian holomorphic vector bundle. Then there exists a unique complex hermitian connection* D. *Let*

$$e = (e_1, \ldots, e_r)$$

be a holomorphic frame, and

$$h_{ji} = \langle e_i, e_j \rangle, \qquad h = (h_{ij}) = h_e.$$

We have the following relations for the connection and curvature matrices:

2.2. $\theta_e = h^{-1} \partial h$, *and* θ_e *is of type* $(1, 0)$.

2.3. $\bar{\partial}\theta_e = -\theta_e \wedge \theta_e$.

2.4. $\Theta_e = \bar{\partial}\theta_e$ *is of type* $(1, 1)$, *and*

$$\bar{\partial}\Theta_e = 0, \qquad \partial\Theta_e = [\Theta_e, \theta_e].$$

Proof. We have

$$dh_{ji} = d\langle e_i, e_j \rangle = \langle De_i, e_j \rangle + \langle e_i, De_j \rangle$$
$$= \sum_k h_{jk}\theta_{ki} + \sum_k \bar{\theta}_{kj}h_{ki}.$$

Since $\bar{\partial}e_i = 0$ because e is a holomorphic frame, we know that θ_{ki} is a form of type $(1, 0)$ for all k, i. Then $\bar{\theta}_{kj}$ is of type $(0, 1)$. Since $d = \partial + \bar{\partial}$, equating forms of the same type on the left- and right-hand side we obtain

$$\partial h = h\theta \qquad \text{and} \qquad \bar{\partial}h = {}^t\bar{\theta}h.$$

But h is hermitian, that is ${}^th = \bar{h}$, so that these two equations are equivalent to each other, and to solve them it suffices to solve either one, say the first, by letting

$$\theta = h^{-1}\partial h$$

which gives both the existence and uniqueness of the desired connection, and also shows that it satisfies **2.2.**

For the other identities, we have

$$\partial\theta_e = \partial h^{-1} \wedge \partial h$$

because $\partial\partial = 0$. Since $hh^{-1} = I$ we also have

$$(\partial h)h^{-1} + h\partial h^{-1} = 0 \qquad \text{so} \qquad \partial h^{-1} = -h^{-1}(\partial h)h^{-1}$$

and therefore

$$\partial\theta_e = -h^{-1}(\partial h)h^{-1} \wedge \partial h = -\theta_e \wedge \theta_e$$

as predicted in **2.3**.

Finally, **2.4** follows immediately from **1.5** and **1.6**, that

$$\Theta_e = d\theta_e + \theta_e \wedge \theta_e \qquad \text{and} \qquad d\Theta_e = [\Theta_e, \theta_e],$$

and the fact that $d = \partial + \bar{\partial}$. This concludes the proof.

Example 2.5. The one-dimensional case. Suppose dim $X = 1$, and let z be a complex coordinate, corresponding to a frame e on TX and the frame dz on $T^{\vee}X$. Let

$$\omega = h_0(z)\frac{\sqrt{-1}}{2\pi} dz \wedge d\bar{z}$$

be a positive $(1, 1)$-form, so the function $h = h_0/\pi$ defines the hermitian metric H_ω. We have $h_0 = \pi\langle e, e\rangle$, and

$$\theta_e = h^{-1}\partial h = \partial \log h,$$

2.5.1 $$\Theta_e = \bar{\partial}\theta_e = -\partial\bar{\partial} \log h = 2\pi\sqrt{-1}\, dd^c \log h,$$

so in full coordinates,

2.5.2 $$\frac{1}{2\pi\sqrt{-1}}\Theta_e = \left[\frac{1}{h}\frac{\partial^2 h}{\partial\bar{z}\partial z} - \frac{1}{h^2}\frac{\partial h}{\partial\bar{z}}\frac{\partial h}{\partial z}\right]\frac{\sqrt{-1}}{2\pi} dz \wedge d\bar{z}.$$

Recalling the definition of the Ricci form from Chapter IV, §2 we then obtain the basic formula

2.5.3 $$\Theta_e = 2\pi\sqrt{-1}\,\mathrm{Ric}(\omega),$$

relating the curvature matrix (here a 1×1 matrix of $(1, 1)$-form) with a notion we already had. Since the Griffiths function G_ω was defined in Chapter IV, §2 by

$$\mathrm{Ric}(\omega) = G_\omega\omega,$$

we can write

2.5.4 $$\Theta_e = 2\pi\sqrt{-1}\, G_\omega\omega_e.$$

This expresses the curvature matrix in terms of the original positive $(1, 1)$-form.

Observe again how much more natural the Griffiths function is here, rather than its negative, the Gauss curvature, in case the Ricci form $\text{Ric}(\omega)$ turns out to be positive. This is precisely the case when the manifold turns out to be hyperbolic. Note especially that Θ_e is not real but pure imaginary.

In arbitrary dimension, we define the **Ricci tensor** R_H or R_ω, determined uniquely by the complex hermitian length function H or the positive $(1, 1)$-form ω, to be

$$R_\omega = R_H = \frac{1}{2\pi\sqrt{-1}}\,\Theta \in H^0(\mathscr{A}^{1,1}(\text{Hom}(E, E))).$$

This terminology is compatible with the terminology of the Ricci form in the one-dimensional case. The higher dimensional expression

$$R_{H,e} = \frac{1}{2\pi\sqrt{-1}}\,\bar{\partial}(h^{-1}\partial h)$$

corresponds precisely to our previous one-dimensional expression

$$\text{Ric}(\omega) = \frac{1}{2\pi\sqrt{-1}}\,\bar{\partial}\partial \log h = dd^c \log h.$$

In the higher dimensional case, the operators ∂ and $\bar{\partial}$ came in the natural order. We always have

$$\partial\bar{\partial} = -\bar{\partial}\partial,$$

so the signs are consistent.

The signs are a pain throughout the literature. In his great paper [Gri 1], Griffiths cleared up a number of relationships concerning "positivity" and "negativity" in various contexts, and posed a number of problems, some of which were later solved. At the same time, he corrected a mistake in signs from a previous paper [Gri 0], which had led to one false result. Basically, Griffiths usually takes the position that a given formula is true up to sign (and other factors like powers of 2 and factorials). The sign convention should be determined by philosophical reasons and "must" come out "right". But in practice, it's a pain to check every step. For instance, in Griffiths–Harris, different conventions are adopted from [Gri 1], and some formulas are "wrong", not only philosophically, but leading to contradictions. Among other things, the sign convention concerning positivity on the top of p. 79 (and which I adopt here), is not compatible with the assertion that "curvature decreases in holomorphic subbundles" made later on that page, if "curvature" is defined by that convention. I have adopted the Griffiths convention whereby the signs all point together in the positive direction because it best reflects the formal structure of the objects involved. I give names to these objects like Griffiths function and Ricci function. The corresponding "curvatures" are minus those functions. I hope that the sign conventions which I have adopted are OK. I have made my share of sign mistakes in the past.

Remark on terminology. A **tensor** is a section of some bundle or other cooked up from the tangent bundle and whatever else, like E in the present instance. Thus the terminology adopted here is *in essence* classical. I don't like the word tensor, because *in fact*, classically, tensors are NOT sections of some bundle, but things like

$$R^l_{ijk}$$

with protruding indices, that have always called to my mind the claws on a rather unpleasant prying insect. When these claws are written down without an accompanying explanation of what they represent, let alone a proof that they do represent something geometric, I find them repugnant. Sad to say, in practice, the situation is complicated enough that one is led to use such claws to prove certain results. But as Griffiths once emphasized: "The general philosophy is that terms which have no intrinsic meaning may, by suitable choice of coordinates, be made equal to zero at a given point." When this happens, the formulas become sufficiently simple to be serviceable and to reflect the geometry. One example of such a frame is given by the next lemma, put here for want of a better place. Other such lemmas will be given in the next section as needed. Historically, frames first arose in Cartan, who developed them in a spirit anti-indices not unlike what I have expressed above.

Lemma 2.6. *Let* Θ *be the curvature of a complex hermitian connection. Given a point* $x \in X$, *there exists a holomorphic frame* e *in a neighborhood of this point with local coordinates* z *such that* x *corresponds to* $z = 0$, *and satisfying:*

2.6.1 $h_e(z) = I + O(|z|^2)$,   *that is* $h_e(0) = I$ *and* $dh_e(0) = 0$.

In particular, $\partial h_e(0) = 0$, $\bar{\partial} h_e(0) = 0$. *With respect to such a frame, we have*

2.6.2. $\Theta_e(0) = \bar{\partial}\partial h_e(0)$.

Proof. The second relation follows from the first, since $\theta_e = h^{-1}\partial h$, and $\Theta_e = \bar{\partial}\theta_e$ by **2.4**, so when we evaluate at 0, the terms which still have only a first derivative in them vanish at the origin to leave the expression $\bar{\partial}\partial h_e(0)$.

As to the first relation, we prove it in two steps. First, under a linear change of coordinates by means of a matrix $g \in \mathrm{GL}_r(\mathbf{C})$ the matrix $h_e(0)$ changes to $h_{eg}(0) = g^*h_e(0)g$, where $g^* = {}^t\bar{g}$. Since $h_e(0)$ is hermitian positive definite, we can find g such that $h_{eg}(0) = I$, and therefore, without loss of generality, we can assume that

$$h(z) = I + O(|z|).$$

We shall then make a further change of coordinates by means of a matrix

$$g(z) = I + A(z),$$

where $A(z) = \big(a_{ij}(z)\big)$ is a matrix of holomorphic functions such that $A(0) = O$. In fact we shall see that it suffices to take the functions $a_{ij}(z)$ to be linear in z, and we determine the equation they must satisfy. It suffices that

$$d\big(g^*(z)h(z)g(z)\big) = 0 \qquad \text{at} \quad z = 0.$$

But

$$d\big(g^*(z)h(z)g(z)\big) = dh(z) + dA^*(z) \cdot h(z) + h(z) \cdot dA(z)$$

$$+ \text{ terms containing } A(z) \text{ or } A(z)^* \text{ as factors.}$$

Therefore

$$dh_{eg}(0) = \partial h(0) + \bar{\partial}h(0) + dA^*(0) + dA(0).$$

Now we let

$$a_{ij}(z) = \sum_{k=1}^{n} a_{ijk} z_k,$$

where

$$a_{ijk} = -\frac{\partial h_{ij}}{\partial z_k}(0).$$

Then $\bar{\partial}A(z) = 0$ since $A(z)$ is holomorphic in z, and thus

$$dA(0) = -\partial h(0).$$

Since h is hermitian, this implies

$$dA^*(0) = -\bar{\partial}h(0),$$

whence $dh_{eg}(0) = 0$, thus concluding the proof of the lemma.

Remark. The first part of Lemma 2.6 reproves Lemma 1.10, which says that there exists a frame e at a given point x such that $\theta_e(0) = 0$, because in the present context, $\theta_e = h^{-1}\,\partial h$ by **2.2**, and $\theta_e(z) = O(|z|)$.

V, §3. THE RICCI FUNCTION

Let E be a holomorphic, hermitian vector bundle over the complex manifold X. We discuss some punctual multilinear algebra. Let $x \in X$. Given a (1,1)-form ω there is associated to it a sesquilinear form $\langle\ ,\ \rangle_\omega$. Let $A \in \operatorname{Hom}(E, E)$. We look at these at the given point x, so if necessary we put an index and write ω_x or $\langle\ ,\ \rangle_{\omega,x}$. Then we may form the scalar-valued function

$$(T_x \otimes E_x) \times (T_x \otimes E_x) \to \mathbf{C}$$

such that if $v, v' \in T_x$ and $e, e' \in E_x$ then

$$(v \otimes e) \times (v' \otimes e') \mapsto \langle v, v' \rangle_{\omega_x} \langle A_x e, e' \rangle_{E_x}.$$

This function depends bilinearly on ω_x and A_x, and so our scalar-valued function can also be viewed as a function of a variable in $H^0(\mathscr{A}^{1,1}(\operatorname{Hom}(E, E)))$. Indeed, given $\Gamma \in H^0(\mathscr{A}^{1,1}(\operatorname{Hom}(E, E)))$ then Γ_x can be expressed as a sum of terms

$$\sum \omega_x \otimes A_x, \qquad \text{where} \quad \omega_x \in \textstyle\bigwedge_{\mathbf{C}}^{1,1} T_x^\vee \quad \text{and} \quad A_x \in \operatorname{Hom}_{\mathbf{C}}(E_x, E_x),$$

so Γ induces a scalar-valued function (the **contraction** of Γ)

$$(T \otimes E) \times (T \otimes E) \to \mathbf{C},$$

which we denote by the symbols on decomposable elements

$$\Gamma(v, v'; e, e') \qquad \text{or} \qquad \langle v \otimes e, v' \otimes e' \rangle_\Gamma.$$

Such a tensor Γ may be viewed as a sesquilinear form on $T \otimes E$. We can then apply this general remark to the case when $\Gamma = R_H$ is the Ricci tensor. The scalar product determined by R_H as above on decomposable elements $v \otimes e$ will be called the **Ricci product**. We say that Γ is **Griffiths positive** if for decomposable elements we have

$$\langle v \otimes e, v \otimes e \rangle_\Gamma > 0 \qquad \text{for all} \quad v \neq 0, \quad e \neq 0.$$

This is the notion which is relevant for us at the moment. If the above scalar product is positive for all non-zero elements of $T \otimes E$, then one

says that Γ is **Nakano positive** in light of [Na]. For good book exposi-
tions of Nakano's result, see Wells [We] and Griffiths–Harris. In [Gri 0]
and [Gri 1], Griffiths investigated systematically these two notions, and
made clear the different roles played by *all* elements in the tensor product, and the *decomposable* elements.

Special Case. Suppose $E = TX$ is the tangent bundle, with its given
hermitian length function $H = H_\omega$ corresponding to the positive $(1, 1)$-
form ω. We define the **Ricci function**

$$\mathbf{r}_\omega \text{ or } \mathbf{r}_H \colon PTX \to \mathbf{C}$$

by the formula

$$\mathbf{r}_H(v) = \mathbf{r}_\omega(v) = R_H(v, v; v, v)/|v|_H^4.$$

This is the contraction of R_H on $(v \otimes v, v \otimes v)$ made homogeneous. Thus
in this case, we are able to set all four variables in the scalar product
equal to each other. Our terminology is consistent with the terminology
we adopted in the one-dimensional case:

Proposition 3.1. *Let (X, ω) be a one-dimensional complex hermitian
manifold. Then for any unit vector u in $T_x X$ we have*

$$\mathbf{r}_\omega(u) = G_\omega(x).$$

Proof. This is immediate from the formula giving the definition of the
Ricci product, together with **2.5.4**.

We shall see in Theorem 3.4 that $\mathbf{r}_\omega$ is real valued in general, by relating this function to the Griffiths function in the one-dimensional case,
which we know is real valued from Chapter IV, §2. This is the invariant
way of seeing the reality, but it also comes out from the special coordinate expressions of Lemma 3.2.

Remark on terminology. By definition, the **holomorphic sectional curvature** as in [G–K] and [Ko 4] is *minus* the Ricci function.

Next we shall obtain a result of [Wu 2] relating the Ricci function in
the one-dimensional case and the higher dimensional case. The proof
will depend on normalizing coordinates in a certain way. The next
lemmas give such normalizations.

Lemma 3.2. *Let (X, ω) be a complex hermitian manifold, and let $x \in X$. Let u be a unit vector in $T_x X$. Then there exist complex coordinates $z_1, \ldots, z_n$ on some open neighborhood U of x having the following properties. We let e be the frame of constant, standard unit vectors over U, that is for all z,*

$$e_1(z) = {}^t(1, 0, \ldots, 0), \ldots, e_n(z) = {}^t(0, \ldots, 0, 1).$$

(a)   *x corresponds to the point $z_1 = \cdots = z_n = 0$.*
(b)   *$u = e_n$.*
(c)   *$(e_1, \ldots, e_n)$ is an orthonormal basis at $z = 0$. In other words, if $h(z) = (h_{ij}(z))$ is the matrix with*

$$h_{ij}(z) = \langle e_j, e_i \rangle_z,$$

then $h(0) = I$.

With respect to such coordinates, the Ricci function is given by

$$\mathbf{r}_\omega(u) = \frac{1}{\pi} \left(\frac{\partial^2 h_{nn}}{\partial z_n \, \partial \bar{z}_n} - \sum_{i=1}^n \left| \frac{\partial h_{in}}{\partial z_n} \right|^2 \right)(0).$$

Proof. Start with any complex coordinates such that x is at the origin. After a linear change of coordinates, we can assume that the standard unit vectors form an orthonormal basis at the origin. After a further change of coordinates by a constant unitary matrix, we can also assume that $u = e_n$ is the last element of the frame. This gives (a), (b), (c).

By **2.4** we know that

$$\Theta_e = \bar{\partial}(h^{-1} \, \partial h) = \bar{\partial}(h^{-1}) \wedge \partial h + h^{-1} \, \bar{\partial}\partial h$$

and

$$\Theta_e(0) = -\bar{\partial} h(0) \wedge \partial h(0) + \bar{\partial}\partial h(0).$$

We have

$$\Theta(0) = \sum \Theta_{e, ij}(0) e_j^\vee(0) \otimes e_i(0).$$

The contraction of $e_j^\vee(0) \otimes e_i(0)$ on $u = e_n$ gives 0 unless $i = j = n$, in which case it gives 1. Thus to find $\Theta(u, u, u, u)$ we are reduced to evaluate the hermitian form associated with $\Theta_{e, nn}(0)$ at $u = e_n$. We have

$$\Theta_{e, nn}(0) = \sum_k \partial h_{nk}(0) \wedge \bar{\partial} h_{kn}(0) + \bar{\partial}\partial h_{nn}(0).$$

For any $n \times n$ matrix $C = (c_{ij})$ we note that

$$
{}^t\bar{u}Cu = (0,\ldots,0,1)C\begin{pmatrix} 0 \\ \vdots \\ 0 \\ 1 \end{pmatrix} = c_{nn}.
$$

From this one sees at once that the hermitian form of the nn-component of the sum on the right-hand side gives precisely the sum of absolute values squared in the statement of the lemma. The sign and the constant factor of $1/\pi$ correspond to our normalizations.

To evaluate the other term, we note that

$$
\partial\bar{\partial}h_{nn} = \sum \frac{\partial^2 h_{nn}}{\partial z_i \, \partial \bar{z}_j} \, dz_i \wedge d\bar{z}_j.
$$

We evaluate the associated hermitian form at 0 on the vector $u = e_n$ and we find precisely the other term as stated, with the appropriate constant $1/\pi$. This concludes the proof.

Lemma 3.3. *Let (X, ω) be a complex hermitian manifold and let $x \in X$. Let u be a unit vector in T_xX. Then there exist complex coordinates which, in addition to the properties of Lemma 3.2, also satisfy:*

(d) $$\frac{\partial h_{ij}}{\partial z_n}(0) = \frac{\partial h_{ij}}{\partial \bar{z}_n}(0) = 0 \qquad \text{for all } i, j.$$

Under this additional condition, we have

$$
\mathbf{r}_\omega(u) = \frac{1}{\pi} \frac{\partial^2 h_{nn}}{\partial z_n \, \partial \bar{z}_n}(0).
$$

Proof. The second assertion is immediate from the nature of the coordinates, since the extra terms in the expression of Lemma 3.2 vanish at 0. We are reduced to proving the existence of coordinates satisfying (d). We have to fiddle with second-order terms.

Let the coordinates obtained in Lemma 3.2 now be denoted by

$$
w = (w_1,\ldots,w_n).
$$

Let $g = (g_{ij})$ be the matrix of the hermitian form with respect to w, so $g_{ij}(w) = \langle e_j, e_i \rangle_w$. Let

$$
z_i = w_i + \sum_{j,k} a^i_{jk} w_j w_k,
$$

where a_{jk}^i are constants such that

$$a_{nj}^i + a_{jn}^i = \frac{\partial g_{ij}}{\partial w_n}(0) \qquad \text{for} \quad i, j = 1, \ldots, n.$$

Such constants can always be found, there is no subtlety in a solution.

We let $h_{ij}(z) = \langle e_j, e_i \rangle_z$. The properties (a), (b), (c) are satisfied for $h(z) = (h_{ij}(z))$. There remains to check that we have achieved condition (d) concerning the partial derivatives. Since

$$\frac{\partial h_{ij}}{\partial \bar{z}_n} = \overline{\frac{\partial h_{ji}}{\partial z_n}}$$

it suffices to prove that

$$(\partial h_{ij}/\partial z_n)(0) = (\partial h_{ij}/\partial w_n)(0) = 0$$

for all i, j. Here goes, it could be worse.

By the definition of how the hermitian metric changes under changes of charts, we have

$$(*) \qquad\qquad g_{ij}(w) = \sum_{k,l} \overline{\frac{\partial z_k}{\partial w_i}} h_{kl}(z) \frac{\partial z_l}{\partial w_j}.$$

Also we have the identity

$$\frac{\partial^2 z_l}{\partial w_n \, \partial w_j} = \frac{\partial g_{lj}}{\partial w_n}(0)$$

coming from the way we chose the coefficients a_{nj}^i. We take the derivative of expression $(*)$ at 0, namely we compute

$$\frac{\partial g_{ij}}{\partial w_n}(0).$$

There will be three sums, as the partial derivative $\partial/\partial w_n$ moves over the three factors. The bar over the first factor makes it so that the holomorphic derivative vanishes on this anti-holomorphic factor. Then we use orthogonality relations like

$$(\partial z_k/\partial w_i)(0) = \delta_{ki} \qquad \text{and} \qquad h_{il}(0) = \delta_{il}.$$

All but one term vanish in the second sum and in the third sum, and we get

$$\frac{\partial g_{ij}}{\partial w_n}(0) = \frac{\partial h_{ij}}{\partial w_n}(0) + \frac{\partial g_{ij}}{\partial w_n}(0)$$

using the definitions of the a_{ij}^i and a_{jn}^i. This proves that

$$\frac{\partial h_{ij}}{\partial w_n}(0) = 0,$$

and concludes the proof of the lemma.

Theorem 3.4 ([Wu 2]). *Let (X, ω) be a complex hermitian manifold. Let $x \in X$.*

(i) *Given a unit vector $u \in T_x X$, there exists locally at x a complex submanifold Y of dimension 1, tangent to u at x, such that*

$$\mathbf{r}_\omega(u) = \mathbf{r}_{\omega|Y}(u) = G_{\omega|Y}(x).$$

(ii) *For every unit vector $u \in T_x X$ we have*

$$\inf_Y G_{\omega|Y}(x) = \mathbf{r}_\omega(u),$$

where the inf is taken over all complex submanifolds Y of dimension 1, tangent to u at x.

Proof. For the first part, we select the holomorphic coordinates as in Lemma 3.3. Define Y locally in terms of these coordinates by

$$z_1 = \cdots = z_{n-1} = 0.$$

Then by definition of $\mathbf{r}_{\omega|Y}(u) = G_{\omega|Y}(x)$ we find exactly the same expression as that of Lemma 3.3, thus proving (i).

The second assertion is an immediate consequence of Proposition 3.1 and (i), using the following result which says that the Ricci function increases when passing to complex submanifolds.

Proposition 3.5. *Let (X, ω) be a complex hermitian manifold. Let X' be a complex submanifold. Let $x \in X'$ and let u be a unit vector in*

$$T_x X' \subset T_x X.$$

Then

$$\mathbf{r}_\omega(u) \leqq \mathbf{r}_{\omega|X'}(u).$$

Proof. We can choose complex coordinates in a neighborhood of x such that X' is defined by

$$z_1 = \cdots = z_r = 0.$$

so $z_{r+1},\ldots,z_n$ are complex coordinates for X' near x. As in Lemma 3.2, by a linear change of coordinates at the origin, and the fact that the hermitian structure on X' is the induced one, we can achieve that $(e_1,\ldots,e_n)$ form an orthonormal basis at the origin, and $u = e_n$. Then the formula of Lemma 3.2 and the definition of the Ricci functions on X and X' respectively show that

$$\mathbf{r}_\omega(u) = \mathbf{r}_{\omega|X'}(u) - \sum_{i=1}^{r} \left| \frac{\partial h_{in}}{\partial z_n} \right|^2,$$

thus proving the proposition.

Remark. Proposition 3.5 is a very special case of a quite general theorem on vector bundles as in [Gri 1], formula (2.15), p. 198, cf. also Griffiths–Harris, p. 79, where the result is expressed by saying that "*curvature decreases in subbundles*". Because of our choice of signs, we see that *the Ricci function increases in subbundles*.

As far as I can make out, the history of this kind of theorem for various contractions of the Ricci tensor giving rise to scalar "curvature" functions *in the holomorphic case* involves the paper of Griffiths [Gri 1] already cited, expanding [Gri 0], and simultaneously O'Neill [O'N], followed by Goldberg–Kobayashi [G-K].

Using the terminology of the Ricci function, we can rephrase Theorem 3.3 of Chapter IV as follows:

Let (X, ω) be a hermitian complex manifold. If there exists $B > 0$ such that $\mathbf{r}_\omega \geq B$, then X is hyperbolic.

This is in fact the way Kobayashi phrases the result in his book, modulo the terminology (he uses "holomorphic sectional curvature").

If X is compact, it suffices to assume $\mathbf{r}_\omega > 0$, since the existence of B as above then follows by continuity on the unit sphere bundle, which is compact.

We end this section with more properties of the Ricci function. The first formalism is due to Grauert–Reckziegel [G-R].

Let f be a positive C^∞ function defined on an open set in $\mathbf{C}$. We define the associated **Gauss function**

$$G(f) = \frac{1}{f} \frac{\partial^2 \log f}{\partial z \, \partial \bar{z}}.$$

This has the following properties, with h also positive C^∞ defined on the same open set:

GR 1. $G(cf) = c^{-1}G(f)$ for $c > 0$.

GR 2. $fhG(fh) = fG(f) + hG(h)$.

GR 3. $(f + h)^2 G(f + h) \geq f^2 G(f) + h^2 G(h)$.

GR 4. If $G(f) \geq b > 0$ and $G(h) \geq c > 0$, then

$$G(f + h) \geq \frac{bc}{b + c}.$$

Proofs. The first two properties are obvious from the definition. For the third, we use the identity

$$fh(f + h)[(f + h)^2 G(f + h) - f^2 G(f) - h^2 G(h)] = \left| f\frac{\partial h}{\partial z} - h\frac{\partial f}{\partial z} \right|^2 \geq 0,$$

which follows directly from the definition. As to the fourth, we use **GR 3** to get

$$G(f + h) \geq \frac{f^2 G(f) + h^2 G(h)}{(f + h)^2} \geq \frac{f^2 b + h^2 c}{(f + h)^2} \geq \frac{bc}{b + c},$$

this last inequality amounting to trivial algebra (cross multiply). This proves the four Grauert-Reckziegel properties.

We give an application from [Wu 2] for the Ricci function.

Proposition 3.6. *Let X be a complex manifold, and ω_1, ω_2 two positive $(1, 1)$-forms making X hermitian in two ways. Let $\mathbf{r}_1$ and $\mathbf{r}_2$ be the two corresponding Ricci functions. Let $\omega_3 = \omega_1 + \omega_2$. Let $\mathbf{r}_3$ be the Ricci function of ω_3. Then*

$$\mathbf{r}_1 \geq 0 \text{ and } \mathbf{r}_2 \geq 0 \quad \Rightarrow \quad \mathbf{r}_3 \geq 0.$$

Suppose on the other hand that there are constants $b_1 > 0$ and $b_2 > 0$ such that $\mathbf{r}_1 \geq b_1$ and $\mathbf{r}_2 \geq b_2$. Then

$$\mathbf{r}_3 \geq \frac{b_1 b_2}{b_1 + b_2}.$$

Proof. This is immediate from Proposition 3.4, which gives a reduction to the one-dimensional case because of Proposition 3.5. This reduction allows us to apply **GR 4** to conclude the proof.

As a special case of the Grauert-Reckziegel property **GR 1**, and Wu's result, we conclude:

Corollary 3.7. *The hermitian structures on X with positive Ricci functions form a positive cone.*

V, §4. GARRITY'S THEOREM

Kodaira's theorem states that if a line bundle admits a hermitian metric with positive first Chern form, then it is ample. Kobayashi [Ko 5] extended this to vector bundles in terms of a length function. Garrity has proved a simpler criterion, which implies part of Kobayashi's theorem [Ga]. In this section, we prove Garrity's theorem, half of which relies mostly on a previous criterion of Barton [Bar], and a theorem of Kleiman [Kl]. Thus we presuppose some basic facts of algebraic geometry, which we shall list systematically.

Algebraic preliminaries

We follow Barton. By a **variety** we shall mean a projective variety, so a scheme over an algebraically closed field k, reduced, irreducible, and imbeddable in a projective space. A **curve** is a variety of dimension 1.

Let X be a variety over k. Let L be a line bundle over X. By a **curve** Y in X we mean a subvariety of dimension 1. Let $f: Y' \to Y$ be the normalization. We define the **degree** of $L|Y$ and the **intersection number** by

$$(L . Y) = \deg(L|Y) = \deg(f^*L).$$

Note that the degree is defined on a singular curve by reducing the definition to the desingularization. If Y is smooth, then $\deg L$ is defined to be $\deg(D)$, where D is any divisor such that the sheaf of sections of L is isomorphic to $\mathcal{O}_Y(D)$.

We say that L is **numerically equivalent to** 0 if $(L . Y) = 0$ for all curves Y in X. We define:

$N^1(X) =$ group of line bundles modulo numerical equivalence.

$A^1(X) = \mathbf{R} \otimes N^1(X)$.

$N_1(X) =$ free abelian group generated by the curves modulo numerical equivalence.

$A_1(X) = \mathbf{R} \otimes N_1(X)$.

We let $[Y]$, or $[Y]_X$ be the class of Y in $N^1(X)$. The intersection product induces a bilinear product

$$A^1(X) \times A_1(X) \to \mathbf{R}, \qquad (d, z) \mapsto \langle d, z \rangle = (d \cdot z),$$

such that the kernels on each side are 0. By the Néron–Severi theorem, both $A^1(X)$ and $A_1(X)$ are finite dimensional over $\mathbf{R}$, and are dual spaces to each other under this pairing.

Let $f: X' \to X$ be a morphism. Then f induces a contravariant homomorphism

$$f^A: A^1(X) \to A^1(X')$$

by the inverse image map $L \mapsto f^*L$. We also have a covariant homomorphism

$$f_A: A_1(X') \to A_1(X)$$

such that if Y' is a curve in X', then

$$f_A[Y'] = 0 \qquad\qquad \text{if } f(Y') \text{ is a point;}$$

$$f_A[Y'] = (\deg f)[f(Y')] \qquad \text{if } f(Y') \text{ is a curve in } X.$$

The notation $\deg f$ denotes the degree of the function field extension $k(Y')$ over $k(Y)$, where $Y = f(Y')$.

The two maps f^A and f_A are adjoint to each other, namely

$$\langle d, f_A z' \rangle = \langle f^A d, z' \rangle \qquad \text{for} \quad z' \in A_1(X'), \quad d \in A^1(X).$$

By a **curve finite over** X we mean a morphism $f: Y \to X$ of a curve into X such that $f(Y)$ is also a curve, so Y is finite over $f(Y)$. For such a curve, $f_A[Y] \neq 0$ because $(L \cdot f(Y)) \neq 0$ if L is ample on X.

Let E be a vector bundle on X and let $P = \mathbf{P}E$ be its projective bundle of lines. Let

$$\pi: P \to X$$

be the projection. Let $E^\vee$ be the dual bundle. From Grothendieck, one knows that there is a **universal quotient line bundle**

$$\pi^*E^\vee \to Q_P \to 0.$$

Indeed, given morphism $f: Y \to X$ and an exact sequence

$$f^*E^\vee \xrightarrow{\alpha} L \to 0,$$

where L is a line bundle, there exists a unique morphism $f_P: Y \to P$ (over X) such that α is obtained from the universal exact sequence by pullback with f_P^*. Two maps α and β determine the same f_P if and only if $\mathrm{Ker}(\alpha) = \mathrm{Ker}(\beta)$. Cf. [Ha 1], Chapter II, Proposition 7.12.

Let L' be a line bundle on P and suppose $\mathrm{rank}(E) > 1$. Then there exists a unique integer n and line bundle L on X (up to isomorphism) such that

$$L' \approx \pi^* L \otimes Q_P^n.$$

This fact is due to Grothendieck. It follows from the general determination of the K-group as in [F–L], Chapter III, Proposition 2.3 and the fact that $u \mapsto \mathrm{Gr}^1(u - 1)$ is an isomorphism between Pic and Gr^1.

Proposition 4.1. *Let X be a variety. Let E be a vector bundle over X of rank > 1. Let $P = \mathbf{P}E$, and $\pi: P \to X$ the projection. Let $\{d_1, \ldots, d_p\}$ be a basis for $A^1(X)$ and let $\{z_1, \ldots, z_p\}$ be the dual basis with respect to the intersection pairing.*

   (i)   *Let $d_0' \in A^1(P)$ be $d_0' = [Q_P]$ and let $d_i' = \pi^A d_i$ for $i \geq 1$. Then $\{d_0', \ldots, d_p'\}$ is a basis for $A^1(P)$.*

  (ii) *Let $\{z_0', \ldots, z_p'\}$ be the dual basis of $\{d_0', \ldots, d_p'\}$. Then $\pi_A z_i' = z_i$ for $i = 1, \ldots, p$.*

 (iii) *The element z_0' is the class of a line in a fiber. More precisely, let x be a point in X and let Y be a line in the projective space $\pi^{-1}(x) = \mathbf{P}(E(x))$. Then $z_0' = [Y]$.*

Proof. By the fact stated before the proposition, $d_0', \ldots, d_p'$ generate $A^1(P)$. Suppose

$$a_0 d_0' + \cdots + a_p d_p' = 0 \qquad \text{with} \quad a_i \in \mathbf{R}.$$

Let z_0'' be the class of a line in a fiber. Then $\pi_A z_0'' = 0$, and from the adjointness

$$\langle \pi^A d_i, z_0'' \rangle = \langle d_i, \pi_A z_0'' \rangle = 0 \qquad \text{for} \quad i = 1, \ldots, p$$

so $\pi^A d_i = d_i'$ is orthogonal to z_0'' for $i \geq 1$. Hence

$$a_0 \langle d_0', z_0'' \rangle = 0,$$

whence $a_0 = 0$ since d_0' is a hyperplane section on the fiber. Furthermore, since π is surjective, it is a fact that π^A is injective [Kl], IV-4, Corollary 1, whence $a_i = 0$ for $i = 1, \ldots, p$. This proves (i).

As to (ii), we note that

$$\langle d_i, \pi_A z_j' \rangle = \langle \pi^A d_i, z_j' \rangle = \delta_{ij} \qquad \text{for} \quad i, j = 1, \ldots, p,$$

so that $\pi_A z_j'$ and z_j give the same functionals on $A^1(X)$, so $\pi_A z_j' = z_j$. Finally for (iii), for $i = 1, \ldots, p$ we get

$$\langle d_i', z_0'' \rangle = \langle \pi^A d_i, z_0'' \rangle = \langle d_i, \pi_A z_0'' \rangle = 0,$$

while

$$\langle d_0', z_0'' \rangle = 1$$

because d_0' is the class of a hyperplane section on a fiber. Hence $z_0'' = z_0'$ by definition of the dual basis. This concludes the proof of the proposition.

Next we come to criteria for ampleness.

A vector bundle $E^\vee$ is defined to be **ample** if Q_P is ample as a line bundle on $\mathbf{P}E = P$.

In what follows, we shall use norms on the finite dimensional vector space $A_1(X)$. Recall that any two norms on such a space are equivalent, i.e. each is bounded by a constant multiple of the other. One useful norm is the sup norm of the coordinates with respect to a basis as in Proposition 4.1. We let $\| \ \|$ denote such a norm.

If Y is a curve in X we write $\|Y\|$ instead of $\|[Y]\|$ for simplicity. We shall use without proof:

Theorem 4.2 (Kleiman's theorem). *Let X be a non-singular variety over k. Let L be a line bundle on X. Then L is ample if and only if there exists $\varepsilon > 0$ such that for all curves $Y \subset X$ we have*

$$(L.Y) \geq \varepsilon \|Y\|.$$

For the proof see [Ha 3], Chapter I, Theorem 8.1, after [Kl]. Observe that if $f: Y' \to Y$ is a smooth curve over Y, for instance the normalization, then the above inequality can be written

$$\deg(f^*L) \geq \varepsilon \|f_A[Y']\|.$$

Indeed, by passing to finite (ramified) coverings, both sides get multiplied by the degree of the covering.

Theorem 4.3 (Barton [Bar]). *Let X be a non-singular variety over k. Let $\| \ \|$ be a norm on $A_1(X)$. A vector bundle $E^\vee$ on X is ample if and only if there exists $\varepsilon > 0$ such that for every curve $Z \subset P$ not contained in a fiber we have*

$$\deg(Q_P|Z) = (Q_P.Z) \geq \varepsilon \|\pi_A[Z]\|.$$

Proof. Let Z be a curve in P not contained in a fiber. Let

$$[Z] = a_0 z'_0 + \cdots + a_p z'_p \qquad \text{with} \quad a_i \in \mathbf{R}.$$

By Proposition 4.1 we get

$$\pi_A[Z] = a_1 z_1 + \cdots + a_p z_p$$

and

$$a_0 = \langle d'_0, [Z] \rangle = (Q_P . Z) = \deg(Q_P . Z).$$

Suppose first that Q_P is ample. By Kleiman's theorem, using the sup norm, we get

$$
\begin{aligned}
a_0 = (Q_P . Z) &\geqq \varepsilon \|Z\| = \varepsilon \max |a_i| \qquad (i = 0, \ldots, p) \\
&\geqq \varepsilon \max |a_i| \qquad (i = 1, \ldots, p) \\
&= \varepsilon \|\pi_A[Z]\|
\end{aligned}
$$

which proves one implication. Conversely, assume the inequality. Let Z be a curve in P. If Z is contained in a fiber $\pi^{-1}(x)$, then

$$(Q_P . Z) = \deg(Q_P . Z) = \|Z\|.$$

If Z is not contained in a fiber, then Kleiman's criterion and Theorem 4.1 immediately imply that Q_P is ample, thus concluding the proof.

Next we formulate Barton's theorem dually in the form which turns out to be useful. The universal quotient bundle

$$\pi^* E^{\vee} \to Q_P \to 0$$

dualizes to the **universal subbundle**

$$0 \to L_P \to \pi^* E$$

so $L_P = Q_P^{\vee}$, and L_P is the bundle of lines in $\pi^* E$.

Theorem 4.4. *Let X be a non-singular variety over k. Let $\| \ \|$ be a norm on $A_1(X)$. Let $E \to X$ be a vector bundle. Then $E^{\vee}$ is ample if and only if there exists $\varepsilon > 0$ such that for every curve $Z \subset \mathbf{P}E$ not contained in a fiber we have*

$$\deg(L_P | Z) \leqq -\varepsilon \|\pi_A[Z]\|.$$

Proof. Immediate from the fact that $(L_P . Z) = -(Q_P . Z)$.

The differential geometry

We are now through with the algebraic considerations, and we pass to the differential geometry. We follow Garrity [Ga]. Let first X be a complex manifold and let $E \to X$ be a holomorphic bundle. As before we have the universal line subbundle L_P on $P = \mathbf{P}E$.

Let E^* denote the open submanifold of E consisting of the non-zero elements, and similarly for L_P^*. Then we have a natural holomorphic isomorphism

$$\tau \colon E^* \to L_P^*.$$

Indeed, a non-zero vector v spans a line L, which defines a point denoted by $[v]$ in $\mathbf{P}E$. The fiber L can be identified with the fiber of the tautological line bundle L_P at $[v]$. Then $\tau(v)$ is the point on L_P corresponding to $v \in L$.

We have defined the notion of length function in Chapter 0, §1 and required in **LF 3** only a continuity property.

For the rest of this section, we strengthen the notion of length function, and assume in addition that it is smooth (i.e. C^∞) on E^*.

If $\text{rank}(E) = 1$, then such a length function is a hermitian metric. We do not, of course, assume smoothness along the zero section, but for rank 1, it follows automatically.

Proposition 4.5. *The map τ induces a bijection between length functions on E and hermitian metrics on L_P, by means of the formula*

$$|v|_H = |\tau v|_\rho.$$

Proof. This follows at once from the definitions.

We have denoted by ρ the metric on L_P. In terms of the hermitian product on L_P, we have

$$|\tau v|_\rho^2 = \langle \tau v, \tau v \rangle.$$

Now let g be a non-vanishing holomorphic section of E over an open set U. Then g induces a holomorphic section

$$s_g \colon U \to \mathbf{P}E,$$

such that $s_g(x)$ is the natural image of $g(x)$ in $\mathbf{P}E(x)$.

We have a commutative diagram:

The map from E_U^* to P is the canonical map. The map τ_U is a holomorphic isomorphism from E_U^* to its image in L_P^*. We shall expand this diagram.

Instead of putting an index U we also write $E^*|U$ to denote restriction to U. Since τ_U is an isomorphism on its image, there exists a section

$$g_P \colon s_g(U) \to L_P^* | s_g(U),$$

which corresponds to g as on the next diagram. Let $x_0 \in U$. Shrinking U to a smaller neighborhood of x_0 if necessary, there exists a neighborhood V of $s_g(x_0)$ in P and a section

$$G \colon V \to L_P^* | V$$

making the following diagram commutative.

$$
\begin{array}{ccc}
E_U^* & \xrightarrow{\ \tau\ } & L_P^*|s_g(U) \subset L_P^*|V \\[2mm]
{\scriptstyle g}\big\uparrow & & {\scriptstyle g_P}\big\uparrow \qquad\qquad {\scriptstyle G}\big\uparrow \\[2mm]
U & \xrightarrow{\ s_g\ } & s_g(U) \quad \subset \quad V
\end{array}
$$

Indeed, locally $L_P^*|V$ is represented by $V \times \mathbf{C}^*$, and a section is represented by a holomorphic function into $\mathbf{C}^*$. A non-vanishing holomorphic function on $s_g(U)$ for U small enough extends to a non-vanishing holomorphic function on some V.

We call the above diagram the **basic diagram**, which relates the differential geometry locally on X with the differential geometry on P.

In Chapter IV, §2 we defined the **Chern form** $c_1(\rho)$ associated with a hermitian metric on a line bundle. If s is locally a holomorphic section, then locally

$$c_1(\rho) = -dd^c \log |s|^2.$$

Proposition 4.6. *Let g be a non-vanishing holomorphic section of E over U. Then over U,*

$$s_g^* c_1(\rho) = -dd^c \log |g|_H^2.$$

Proof. By definition, $c_1(\rho)$ is represented on the open set V as above by $-dd^c \log |G|^2$. Now:

$$
\begin{aligned}
s_g^* dd^c \log |G|_\rho^2 &= dd^c s_g^* \log |G|^2 \\
&= dd^c \log |G \circ s_g|^2 \\
&= dd^c \log |g_P \circ s_g|^2 \\
&= dd^c \log |\tau \circ g|^2 \\
&= dd^c \log |g|_H^2,
\end{aligned}
$$

which proves the theorem. Note that all absolute value signs except the last should have ρ as an index, whereas the last has H as an index.

We observe that the formulas in the proposition define a real $(1, 1)$-form on U, which we shall denote by

$$c_{1,g}(H) \quad \text{or} \quad c_g(H) \quad \text{for short,}$$

and which we call the **Chern form** of H determined by the section g. It depends both on the length function H and the section g.

Let E be a vector bundle on a compact complex manifold X. Recall that $E^\vee$ is ample if and only if the universal quotient bundle $Q_{\mathbf{P}E}$ is ample.

Theorem 4.7 (Garrity). *Let (X, ω) be a complex projective hermitian manifold, and let $E \to X$ be a holomorphic vector bundle. The following conditions are equivalent.*

(a)   *The dual bundle $E^\vee$ is ample.*

(b)   *There exists a length function H on E and $\varepsilon > 0$ such that for all non-vanishing holomorphic sections g on an open set U we have*

$$c_g(H) \leqq -\varepsilon\omega \quad \text{on} \quad U.$$

Remark. The choice of positive $(1, 1)$-form in the statement of the theorem is irrelevant, because given two positive $(1, 1)$-forms on a compact manifold, each is less than a constant multiple of the other.

Proof. Assume that $E^\vee$ is ample. We use the fact that if a line bundle L is ample, then there exists a hermitian metric ρ on L such that $c_1(\rho) > 0$. One can take the pull-back of the Fubini–Study metric from projective space under a projective imbedding. Hence by assumption, there exists a hermitian metric ρ on L_P such that $c_1(\rho)$ is a negative $(1, 1)$-form. We let H be the corresponding length function on E. By compactness, there exists $\varepsilon > 0$ such that

$$c_1(\rho) + \varepsilon\pi^*\omega < 0.$$

Pulling back by s_g which is a section of π so that $s_g^* \pi^* = \text{id}$, we get

$$s_g^* c_1(\rho) + \varepsilon \omega \leqq 0.$$

This proves condition (b). Observe that in this part of the proof we have not used either Kleiman's or Barton's theorem.

We now come to the converse, and have to make some comments how we are going to apply Barton's theorem.

If Y is a curve and Y_0 is the complement of a finite set of points containing all singular points, and if φ is a $(1, 1)$-form on Y_0 then we **define**

$$\int_Y \varphi = \int_{Y_0} \varphi.$$

This value is independent of the number of points deleted. In particular, it is also equal to the integral of the pull-back of the form to the desingularization. The following is standard elementary.

Proposition 4.8. *Let L be a line bundle on a non-singular curve Y, with a metric ρ. Then*

$$\deg L = \int_Y c_1(\rho).$$

Proof. (Reproduced for the convenience of the reader.) Let s be a rational section of L locally, so

$$c_1(\rho) = -dd^c \log |s|^2.$$

At each point y where s has a zero or pole, we put a small circle $C(y, \varepsilon)$ of radius ε with respect to some choice of holomorphic coordinate, and we apply Stokes' theorem to the complement of these discs. Then

$$\int_Y c_1(\rho) = \lim_{\varepsilon \to 0} \sum_{y \in Y} \int_{C(y, \varepsilon)} d^c \log |s|^2.$$

At a point y, we can represent $|s|^2 = \bar{f} f h$ where f is a meromorphic function at y and h is smooth positive. Then

$$(\partial - \bar{\partial}) \log |s|^2 = (\partial - \bar{\partial}) \log f + (\partial - \bar{\partial}) \log \bar{f} + (\partial - \bar{\partial}) \log h.$$

But $(\partial - \bar{\partial}) \log h$ involves $h^{-1} \partial h$ or $h^{-1} \partial \bar{h}$ which is bounded locally, and so the integral of this term tends to 0 as $\varepsilon \to 0$. Also $\bar{\partial} \log f = 0$ and

$\partial \log \bar{f} = 0$. Hence the integral at y except for a constant factor amounts to

$$\int_{C(y, \varepsilon)} \partial \log f = \int_{C(y, \varepsilon)} f'/f(z) \, dz = 2\pi \sqrt{-1} \, m(y),$$

where $m(y)$ is the multiplicity of s at the point y. One verifies at once that the constant factors come out precisely so as to make the formula of the proposition come out right, as desired.

We come to the proof proper of the converse. Let $\omega = c_1(\sigma)$ where σ is a metric on an ample line bundle L on X. Thus $\omega > 0$. Let $E \to X$ be a holomorphic vector bundle, and *assume* (b), *that there exists a length function H and $\varepsilon > 0$ such that*

$$c_g(H) \leq -\varepsilon\omega \quad \text{for all local sections } g.$$

It suffices to show that Barton's criterion of Theorem 4.4 is satisfied. Let $Z \subset P$ be a curve in P not contained in a fiber. Let

$$Z_0 = Z - \{\text{singular points of } Z \text{ and ramification points of } \pi|Z\}.$$

The integral of a $(1, 1)$-form over Z will be viewed as the integral over Z_0.

Lemma 4.9. *Under assumption* (b), *we have*

$$c_1(\rho) \leq -\varepsilon\pi^*\omega \quad \text{on} \quad Z_0.$$

Proof. The assertion is local on Z_0, so we can restrict our attention to an open subset of Z_0 where π is an isomorphism. Such an open subset can be expressed as $V \cap Z_0$ where V is open in P, and there is a holomorphic section g of E on $\pi(V)$ such that

$$s_g(\pi(V \cap Z_0)) = V \cap Z_0.$$

In other words, π and s_g are inverse maps from $V \cap Z_0$ and its image under π. From the definition of $c_g(H)$ in Proposition 4.6 and the fact that inequalities are preserved under pull-backs, we get

$$c_g(H) = s_g^* c_1(\rho) \leq -\varepsilon\omega \quad \text{locally on } X$$

$$\Rightarrow \quad s_g^* c_1(\rho) \leq -\varepsilon\omega \quad \text{restricted to } \pi(V \cap Z_0)$$

$$\Rightarrow \quad c_1(\rho) \leq -\varepsilon\pi^*\omega \quad \text{restricted to } V \cap Z_0.$$

This proves the lemma.

Finally, we have

$$\deg(L_P | Z) = \int_Z c_1(\rho) \qquad \text{by Proposition 4.8}$$

$$= \int_{Z_0} c_1(\rho) \qquad \text{by definition}$$

$$\leqq -\varepsilon \int_{Z_0} \pi^* \omega \qquad \text{by Lemma 4.9}$$

$$= -\varepsilon n \int_{\pi(Z)} \omega \qquad \text{where } n = [Z : \pi(Z)]$$

$$= -\varepsilon n (L . \pi(Z)) \qquad \text{by Proposition 4.8}$$

$$\leqq -\varepsilon_1 \|\pi_A[Z]\| \qquad \text{by Kleiman's theorem.}$$

Applying Barton's criterion of Theorem 4.4 concludes the proof of Garrity's theorem.

At this point we have all the tools necessary to give the differential geometric proof of Kobayashi's theorem:

Theorem 4.10. *Let X be a projective non-singular variety over* **C.** *If $T^\vee X$ is ample, then X is hyperbolic.*

One goes through the rigmarole with the Schwarz lemma, cf. also the arrangement of Griffiths reproduced in [La 3]. We leave this as an exercise to the reader, especially since we gave Urata's proof earlier.

It is part of the foundations of the theory of ample vector bundles that $T^\vee X$ ample implies the canonical bundle ample. In trying to fit hyperbolically into this, one then has the conjecture:

Conjecture 4.11. *If X is a projective non-singular variety which is hyperbolic, then the canonical class K_X is ample.*

The example of Fermat hypersurfaces of degree d in $\mathbf{P}^n$ with $d \geqq n + 2$ shows that ample canonical class does not imply hyperbolic since such hypersurfaces contain lines. Cf. the discussion in [La 3], and also the discussion following Theorem 3.3 of Chapter IV concerning the possibility of characterizing hyperbolicity by a differential geometric condition. This is one of the goals for which this chapter and the preceding one serve as an introduction.

CHAPTER VI

Nevanlinna Theory

In classical estimates of orders of growth of an entire function, one uses the measure of growth given by

$$M_f(R) = \log \sup_{|z| = R} |f(z)| = \log \|f\|_R.$$

This obviously does not work for meromorphic functions, which represent holomorphic maps of $\mathbf{C}$ into projective space $\mathbf{P}^1$. Nevanlinna showed how to measure growth by another function. Starting from his version of the Poisson–Jensen formula, he was then able to derive a much more subtle growth estimate for meromorphic functions in what he called the Second Main Theorem. This chapter develops Nevanlinna's theory, following the treatment in his book.

For the applications at the beginning of Chapter VII, we do not need all of §3 and §4. We need only Lemma 3.2 and Lemma 3.7. An interested reader can skip immediately from these lemmas to the applications. The proofs of these lemmas occupy only a few lines.

VI, §1. THE POISSON–JENSEN FORMULA

We pick up where elementary courses in complex analysis leave off, but we review quickly the proof of this basic formula before pushing further.

Theorem 1.1 (Poisson). *Let f be holomorphic on the closed disc $\bar{\mathbf{D}}_R$. Let z be inside the disc, and write $z = re^{i\varphi}$. Then*

$$f(z) = \int_0^{2\pi} f(Re^{i\theta}) \, \mathrm{Re} \, \frac{Re^{i\theta} + z}{Re^{i\theta} - z} \frac{d\theta}{2\pi}$$

$$= \int_0^{2\pi} f(Re^{i\theta}) \frac{R^2 - r^2}{R^2 - 2Rr\cos(\theta - \varphi) + r^2} \frac{d\theta}{2\pi}.$$

Proof. By Cauchy's theorem,

$$f(z) = \frac{1}{2\pi i} \int_{S_R} \frac{f(\zeta)}{\zeta - z} \, d\zeta = \int_0^{2\pi} f(Re^{i\theta}) \frac{Re^{i\theta}}{Re^{i\theta} - re^{i\varphi}} \frac{d\theta}{2\pi}.$$

On the other hand, let $z' = R^2/\bar{z} = (R^2/r)e^{i\varphi}$. Then $\zeta \mapsto f(\zeta)/(\zeta - z')$ is holomorphic on $\bar{\mathbf{D}}_R$ so

$$0 = \frac{1}{2\pi i} \int_{S_R} \frac{f(\zeta)}{\zeta - z'} \, d\zeta = \int_0^{2\pi} f(Re^{i\theta}) \frac{re^{i\theta}}{re^{i\theta} - Re^{i\varphi}} \frac{d\theta}{2\pi}.$$

Subtracting yields

$$f(z) = \int_0^{2\pi} f(Re^{i\theta}) \left[\frac{Re^{i\theta}}{Re^{i\theta} - re^{i\varphi}} - \frac{re^{i\theta}}{re^{i\theta} - Re^{i\varphi}} \right] \frac{d\theta}{2\pi}$$

$$= \int_0^{2\pi} f(Re^{i\theta}) \frac{R^2 - r^2}{R^2 - 2Rr\cos(\theta - \varphi) + r^2} \frac{d\theta}{2\pi}.$$

Computing the real part as prescribed in the first formula of the theorem concludes the proof.

If we decompose

$$f = u + iv \qquad \text{with} \quad u = \mathrm{Re}\, f$$

into its real and imaginary part, then the integral expression of the theorem gives separately an integral expression for both u and v. For the real part, we thus get:

(1.2) $$u(z) = \int_0^{2\pi} u(Re^{i\theta}) \, \mathrm{Re} \, \frac{Re^{i\theta} + z}{Re^{i\theta} - z} \frac{d\theta}{2\pi},$$

and also as a corollary, for f itself:

(1.3) $$f(z) = \int_0^{2\pi} u(Re^{i\theta}) \frac{Re^{i\theta} + z}{Re^{i\theta} - z} \frac{d\theta}{2\pi} + iC \qquad \text{(with } C \text{ real constant).}$$

This follows because the function on the right-hand side is holomorphic in z, and its real part coincides with the real part of f. If two holomorphic functions on a connected open set have the same real part, then their difference is a pure imaginary constant, because a non-constant holomorphic function is an open mapping.

In fact, formula (1.3) can be used to prove:

Let u be a real-valued continuous function on $\mathbf{S}_R$. Then there exists a unique function on the closed disc $\bar{\mathbf{D}}_R$ which is continuous harmonic on the interior, and equals u on the boundary.

There exists an analytic function f on $\mathbf{D}_R$ of which u is the real part, and f is uniquely determined up to a pure imaginary constant.

In addition to the above arguments, the proof of the first statement needs an additional approximation technique of Dirac families, cf. my *Complex Analysis*, Chapter VIII, §4 and §5. It is unnecessary for the applications we have in mind.

Now let f be *meromorphic* on $\bar{\mathbf{D}}_R$ and let us write

$$f(z) = f_0(z) \prod_i (z - a_i) \prod_j \frac{1}{(z - b_j)},$$

where f_0 has no zero or pole in the disc $\bar{\mathbf{D}}_R$, and a_i, b_j are the zeros and poles, indexed to take account of their multiplicities. Then there exists a holomorphic function $\log f_0(z)$ on $\mathbf{D}_R$, unique up to an integral multiple of $2\pi\sqrt{-1}$. Note that we now allow f to have zeros or poles on the circle $\mathbf{S}_R$.

In fact the above factorization is not useful to study functions on discs, and it is better to use **Blaschke products**, or as we shall also say, **canonical products** as follows. Given $a \in \mathbf{D}_R$ consider the function

$$G_R(z, a) = G_{R,a}(z) = \frac{R^2 - \bar{a}z}{R(z - a)}.$$

Then $G_{R,a}$ has precisely one pole in $\bar{\mathbf{D}}_R$ and no zeros. We have

$$|G_{R,a}(z)| = 1 \quad \text{for} \quad |z| = R.$$

Given any set S of points a with $|a| < R$ we may form the canonical product

$$G_{R,S}(z) = \prod_{a \in S} G_{R,a}(z).$$

Then again,

$$|G_{R,S}(z)| = 1 \qquad \text{for} \quad |z| = R.$$

Instead of a set of points, we may consider finite families, i.e. finite sequences. In particular, given a meromorphic function f on $\bar{\mathbf{D}}_R$ we can associate with f **two canonical products** corresponding to its zeros and poles, namely:

$$G_{R,f}^0(z) = \prod_{\substack{a \in \mathbf{D}_R \\ f(a) = 0}} \left(\frac{R^2 - \bar{a}z}{R(z - a)}\right)^{\mathrm{ord}_a f}$$

and

$$G_{R,f}^\infty(z) = \prod_{\substack{a \in \mathbf{D}_R \\ f(a) = \infty}} \left(\frac{R^2 - \bar{a}z}{R(z - a)}\right)^{-\mathrm{ord}_a f},$$

where $\mathrm{ord}_a(f)$ is the order of f at a. If a is a pole of f, then a has negative order. Thus the exponents are positive in both products.

Theorem 1.4 (Poisson-Jensen's formula). *Let f be meromorphic on $\bar{\mathbf{D}}_R$ as above. Then for any simply connected open subset of $\mathbf{D}_R$ not containing the zeros or poles of f, there is a real constant C such that for z in this open set we have*

$$\log f(z) = \int_0^{2\pi} \log |f(Re^{i\theta})| \frac{Re^{i\theta} + z}{Re^{i\theta} - z} \frac{d\theta}{2\pi} - \sum_{a \in \mathbf{D}_R} (\mathrm{ord}_a \, f) \log G_R(z, a) + iC.$$

The constant C depends on a fixed determination of the logs.

Proof. Suppose first f has no zeros or poles on $\mathbf{S}_R$. Let

$$h(z) = f(z) \prod \left(\frac{R^2 - \bar{a}z}{R(z - a)}\right)^{\mathrm{ord}_a f}$$

We apply (1.3) to $\log h(x)$. We then use the fact that $|f(z)| = |h(z)|$ when $|z| = R$. We select a determination of the logarithm, which is a homomorphism, except for the pure imaginary constant, to conclude the proof.

Now consider the case when f may have zeros and poles on the boundary. Let $\{R_n\}$ be a sequence of radii having R as a limit. For R_n sufficiently close to R, the zeros and poles of f inside the disc of radius

R_n are the same as the zeros and poles of f inside the disc of radius R, except for the zeros and poles lying on the circle $\mathbf{S}_R$. The left-hand side of the formula is independent of R_n. Let

$$\varphi_n(\theta) = \log|f(R_n e^{i\theta})| \qquad \text{and} \qquad \varphi(\theta) = \log|f(Re^{i\theta})|.$$

There exists a constant $K > 0$ such that $|\varphi_n(\theta)| \leq K|\varphi(\theta)|$ for all n sufficiently large and all θ. Indeed, outside fixed small discs around the possible zeros and poles of f on the circle $\mathbf{S}_R$ this is obvious by continuity, and inside such discs, we use $f(\zeta) \sim (\zeta - \zeta_0)^k$ and draw a picture to verify such an inequality. Since φ is integrable, we can apply the dominated convergence theorem and take the limit as $n \to \infty$, so $R_n \to R$, to conclude the proof of the theorem.

The Poisson–Jensen formula as stated above is actually due to Nevanlinna, who baptized it in that manner.

Suppose that $f(0) \neq 0, \infty$. Then we can evaluate the constant C from the special value $z = 0$, so

$$C = \arg f(0) - \sum (\operatorname{ord}_a f) \arg(-R/a) + 2n\pi$$

with some integer n, and C depends only on the a_i, b_j, $f(0)$ up to a fixed integral multiple of 2π.

As a corollary of Theorem 1.4 we get

For all $z \in \mathbf{D}_R$ which are not zeros or poles of f, we have

(1.5)

$$\log|f(z)| = \int_0^{2\pi} \log|f(Re^{i\theta})| \operatorname{Re} \frac{Re^{i\theta} + z}{Re^{i\theta} - z} \frac{d\theta}{2\pi} - \sum_{a \in \mathbf{D}_R} (\operatorname{ord}_a f) \log|G_R(z, a)|.$$

Suppose 0 is not a zero or pole of f. Then

$$\log|f(0)| = \int_0^{2\pi} \log|f(Re^{i\theta})| \frac{d\theta}{2\pi} - \sum_{\substack{a \in \mathbf{D}_R \\ a \neq 0}} (\operatorname{ord}_a f) \log\left|\frac{R}{a}\right|.$$

In general, let $f(z) = cz^m + \ldots$ with $c \neq 0$ and $m \in \mathbf{Z}$, where $c = c_f$ is the leading coefficient. Then we get the classical **Jensen's formula**

(1.6) $\quad \log|c_f| = \displaystyle\int_0^{2\pi} \log|f(Re^{i\theta})| \frac{d\theta}{2\pi} - \sum_{a \neq 0} (\operatorname{ord}_a f) \log|R/a| - m \log R.$

This follows by applying the previous formula to the function $f(z)/z^m$.

Vojta has pointed out that Jensen's formula is the analogue of the Artin–Whaples product formula (sum formula) in algebraic number theory. Let F be a field. We index absolute values on F by symbols like v, and we write $\| \ \|_v$ for the absolute value, suitably normalized. We define the associated **valuation** by

$$v(a) = -\log \|a\|_v \quad \text{for} \quad a \in F, \quad a \neq 0.$$

Given a family of absolute values on F we say that they satisfy the **product formula** if

$$\prod_v \|a\|_v = 1 \quad \text{that is} \quad \sum_v v(a) = 0.$$

Now fix R and let $F = \mathrm{Mer}(\bar{\mathbf{D}}_R)$ be the field of meromorphic functions on the closed disc (i.e. on an open neighborhood of the closed disc). For each θ we have an absolute value

$$f \mapsto |f(Re^{i\theta})| = \|f\|_{\theta, R}.$$

The standard definition of an absolute value has to be extended to allow ∞ as a value, since f may have poles on the circle. Then the corresponding valuation is given by

$$v_{\theta, R}(f) = v_\theta(f) = -\log |f(Re^{i\theta})|.$$

We think of $v_\theta(f)$ as the "order of f at θ". At each point $a \in \mathbf{D}_R$ we have a valuation given by

$$v_{a, R}(f) = v_a(f) = (\mathrm{ord}_a f) \log \left| \frac{R}{a} \right| \quad \text{if} \quad a \neq 0,$$

$$v_{a, R}(f) = v_a(f) = (\mathrm{ord}_a f) \log R \quad \text{if} \quad a = 0.$$

The extra factors involving $\log |R/a|$ and $\log R$ are weighting factors, similar to $\log p$ in number theory, when p is a prime number. We may express Jensen's formula in a form which is a slight perturbation of the product formula, namely

$$\boxed{\int_0^{2\pi} v_\theta(f) \frac{d\theta}{2\pi} + \sum_{a \in \mathbf{D}_R} v_a(f) = -\log |c_f|.}$$

Note that $f \mapsto -\log|c_f|$, although multiplicative, is not a valuation because the leading coefficient of a sum has usually nothing to do with the leading coefficient of each term.

The sums over zeros and poles are going to occur frequently, so we shall give them names as follows. Let $a \in \mathbf{C}$. We let:

$n_f(R, 0) = $ number of zeros of f in $\bar{\mathbf{D}}_R$;

$n_f(R, \infty) = $ number of poles of f in $\bar{\mathbf{D}}_R$

$\qquad\qquad = n_f(R).$

Thus $n_f(R, 0) = n_{1/f}(R, \infty) = n_{1/f}(R)$.

Remark. The numbers of zeros and poles are counted positively. Thus

$$n_f(0, 0) - n_f(0, \infty) = m = \text{order of } f \text{ at } 0.$$

For these definitions, we allow the degenerate case $R = 0$. We let the **counting function** be:

$$N_f(R, 0) = \sum_{\substack{a \in \mathbf{D}_R \\ a \neq 0, \, f(a) = 0}} (\operatorname{ord}_a f) \log \left| \frac{R}{a} \right| + n_f(0, 0) \log R$$

$$= \sum_{\substack{a \in \mathbf{D}_R \\ \operatorname{ord}_a(f) > 0}} v_a(f),$$

$$N_f(R, \infty) = \sum_{\substack{f(a) = \infty \\ a \neq 0}} -(\operatorname{ord}_a f) \log \left| \frac{R}{a} \right| + n_f(0, \infty) \log R$$

$$= N_f(R) = \sum_{\substack{a \in \mathbf{D}_R \\ \operatorname{ord}_a(f) < 0}} -v_a(f).$$

We rewrite Jensen's formula with the above notation.

Theorem 1.7 (Jensen's Formula at $z = 0$).

$$\log|c_f| + N_f(R, 0) = \int_0^{2\pi} \log|f(Re^{i\theta})| \frac{d\theta}{2\pi} + N_f(R, \infty).$$

It will be useful to express $N_f(R)$ as an integral.

Proposition 1.8. *We have*

$$N_f(R) = \int_0^R [n_f(t) - n_f(0)] \frac{dt}{t} + n_f(0) \log R.$$

Proof. The function

$$t \mapsto n_f(t) - n_f(0)$$

is a step function, which is equal to 0 for t sufficiently close to 0. If we decompose the interval $[0, R]$ into subintervals whose end points are the jumps in the absolute values of the zeros of f, integrate over each such interval where the integrand is constant, and take the sum, then the formula of the proposition drops out.

The following special case of Theorem 1.7 is worth recording separately. Because it is so simple, we give a proof for it *ab ovo*. For $\alpha > 0$ we let

$$\log^+ \alpha = \max(0, \log \alpha).$$

Lemma 1.9. *Let $b \in \mathbf{C}$. Then*

$$\int_0^{2\pi} \log |b - e^{i\theta}| \frac{d\theta}{2\pi} = \log^+ |b|.$$

Proof. If $|b| > 1$ then $\log |b - z|$ for $|z| < 1 + \varepsilon$ is harmonic, and $\log^+ |b| = \log |b|$, so the formula is true by the mean value property for harmonic functions. If $|b| < 1$, then

$$\int_0^{2\pi} \log |b - e^{i\theta}| \frac{d\theta}{2\pi} = \int_0^{2\pi} \log |be^{-i\theta} - 1| \frac{d\theta}{2\pi}$$
$$= \log |-1| = 0 = \log^+ |b|.$$

If $|b| = 1$ the formula follows by continuity of each side in b, and the absolute convergence of the integral on the left. This proves the lemma.

We return to the canonical products which give explicit examples of Jensen's formula. For $a \in \mathbf{D}_R$ recall that

$$G_{R,a}(z) = G_a(z) = \frac{R^2 - \bar{a}z}{R(z - a)} \quad \text{and} \quad |G_a(z)| = 1 \quad \text{for} \quad |z| = R.$$

Let

$$G(z) = \prod_a G_{R,a}(z),$$

where the product is taken over some finite family of $a \in \mathbf{D}_R$, possibly with multiplicities. From the maximum modulus principle applied to $1/G$ we conclude that

$$|G(z)| \geq 1 \qquad \text{for} \quad |z| \leq R$$

and

$$\log|G(z)| = \log^+|G(z)| \qquad \text{for} \quad |z| \leq R.$$

For any meromorphic function f we define

$$m_f(r) = \int_0^{2\pi} \log^+|f(re^{i\theta})| \frac{d\theta}{2\pi}.$$

Such integrals will occur all the time in the sequel. Then we get:

Proposition 1.10. *For $r < R$ and $G_a = G_{R,a}$ we have*

$$m_{G_a}(r) = \int_0^{2\pi} \log^+|G_a(re^{i\theta})| \frac{d\theta}{2\pi} = \log\frac{R}{r} - \log^+\left|\frac{a}{r}\right|.$$

This is immediate from Lemma 1.9. We replace $\log^+$ by $\log$, and use the value given in this proposition for each term. Alternatively, it is also an immediate consequence of Jensen's formula in Theorem 1.7. We have

$$\log^+|G(z)| = \log|G(z)| = \sum_a \log^+|G_a(z)|.$$

Therefore we obtain the same additivity relation for m_G, that is if G_1, G_2 are canonical products, then

$$m_{G_1 G_2} = m_{G_1} + m_{G_2}.$$

We then get the analogous expression for the full product G.

Proposition 1.11. *For $r < R$ and $G = G_{R,f}^\infty$ we have*

$$m_G(r) = \int_0^{2\pi} \log^+|G_{R,f}^\infty(re^{i\theta})| \frac{d\theta}{2\pi} = N_f(R, \infty) - N_f(r, \infty),$$

and similarly with 0 replacing ∞.

Proof. Either apply Proposition 1.10 and use the definitions, or use Jensen's formula of Theorem 1.7, and compute c_G for $G = G_{R,f}^\infty$. The answer drops out.

VI, §2. NEVANLINNA HEIGHT AND THE FIRST MAIN THEOREM

We reformulate Jensen's formula in terms of other functions as in Nevanlinna. Let α be a positive real number. From the definition

$$\log^+ \alpha = \log \max(1, \alpha) = \max(0, \log \alpha),$$

we find

$$\log \alpha = \log^+ \alpha - \log^+ 1/\alpha \qquad \text{and} \qquad |\log \alpha| = \log^+ \alpha + \log^+ 1/\alpha.$$

We define:

$$m_f(R, \infty) = \int_0^{2\pi} \log^+ |f(Re^{i\theta})| \frac{d\theta}{2\pi},$$

$$T_f(R, \infty) = T_f(R) = m_f(R, \infty) + N_f(R, \infty).$$

The function T_f is usually called the **characteristic function**. But in light of Vojta's analogy between Nevanlinna theory and the theory of heights in number theory, we shall call T_f the **height function**, or **Nevanlinna height function** after we define another later. The function m_f is called the **proximity function**. It measures how close f comes to a given complex number, or to infinity.

We apply the definitions to Jensen's formula, which we rewrite

$$\log|c_f| + N_f(R, 0) = \int_0^{2\pi} \log^+ |f(Re^{i\theta})| \frac{d\theta}{2\pi}$$

$$- \int_0^{2\pi} \log^+ \frac{1}{|f(Re^{i\theta})|} \frac{d\theta}{2\pi} + N_f(R, \infty).$$

Then in terms of the newly defined functions m_f, the formula becomes

$$\log|c_f| + N_{1/f}(R, \infty) + m_{1/f}(R, \infty) = N_f(R, \infty) + m_f(R, \infty);$$

or using the definition of T_f:

Proposition 2.1. $\log |c_f| + T_{1/f} = T_f$.

So far we have had exact formulas. We shall now derive approximate formulas, modulo $O_f(1)$ where the bounded term $O_f(1)$ depends on the function. This dependence will be rather simple in most cases. Using an $O(1)$ term, the formulas will exhibit an interesting formalism. We base estimates on the easy inequalities

$$\log^+(\alpha_1 \ldots \alpha_n) \leq \sum_{k=1}^{n} \log^+ \alpha_k,$$

$$\log^+(\alpha_1 + \cdots + \alpha_n) \leq \max \log^+ \alpha_i + \log n.$$

For the proof, look at $\log^+(n \cdot \max \alpha_i)$. The second inequality will be used in weaker form

$$\log^+(\alpha_1 + \cdots + \alpha_n) \leq \sum \log^+ \alpha_i + \log n.$$

Proposition 2.2. *For f meromorphic on $\bar{\mathbf{D}}_R$, and any complex number a, we have for $0 \leq r \leq R$:*

$$T_f(r) = T_{f-a}(r) + O_a(1),$$

where $|O_a(1)| \leq \log^+ |a| + \log 2$.

Proof. Apply Jensen's formula to $f - a$. We find

$$\log |c_{f-a}| + N_{f-a}(r, 0) = \int_0^{2\pi} \log |f(re^{i\theta}) - a| \frac{d\theta}{2\pi} + N_f(r).$$

Now use

$$\log |f(re^{i\theta}) - a| = \log^+ |f(re^{i\theta}) - a| - \log^+ \frac{1}{|f(re^{i\theta}) - a|}.$$

Then we obtain

$$\log |c_{f-a}| + N_{1/(f-a)}(r) + m_{1/(f-a)}(r) = \int_0^{2\pi} \log^+ |f(re^{i\theta}) - a| \frac{d\theta}{2\pi} + N_f(r).$$

Finally use **2.1** and the inequality $\log^+(\alpha \pm \beta) \leq \log^+ \alpha + \log^+ \beta + \log 2$, both for $f(re^{i\theta}) - a$ and for $f(re^{i\theta}) - a + a$ to conclude the proof.

Propositions 2.1 and 2.2 together are often called the **First Main Theorem** of Nevanlinna theory.

For the record, we now list inequalities for the various functions, arising from the corresponding inequalities for $\log^+$.

Let h_f denote T_f, or m_f or N_f. Then

$$h_{fg} \leqq h_f + h_g,$$

$$h_{f_1 + \cdots + f_n} \leqq h_{f_1} + \cdots + h_{f_n} + \log n,$$

$$N_{f_1 + \cdots + f_n} \leqq N_{f_1} + \cdots + N_{f_n}.$$

The inequalities for T_f follow from those for m_f and N_f separately. Note that for $N_f(r, \infty)$, the inequality for a sum $f = f_1 + \cdots + f_n$ has no extra constant coming out. The extra $\log n$ for the inequality with T_f arises in the inequality for m_f.

Next we prove some properties of T_f. By convention, we let

$$\log 0 = -\infty \qquad \text{so} \qquad \log^+ 0 = 0.$$

Also for the following theorem, it is convenient to use the notation

$$N_f(r, a) = N_{f-a}(r, 0) \qquad \text{and} \qquad n_f(r, a) = n_{f-a}(r, 0).$$

Theorem 2.3 (Cartan). *If $f(0) \neq \infty$ then*

$$T_f(r) = \int_0^{2\pi} N_f(r, e^{i\theta}) \frac{d\theta}{2\pi} + \log^+ |f(0)|$$

$$= \int_0^r \int_0^{2\pi} n_f(t, e^{i\theta}) \frac{d\theta}{2\pi} \frac{dt}{t} + \log^+ |f(0)|,$$

and if $f(0) = \infty$ then replace $\log^+|f(0)|$ by $\log |c_f|$. In particular, T_f is an increasing function of r.

Proof. Suppose first $f(0) \neq \infty$. For each θ we apply Jensen's formula to the function $f(z) - e^{i\theta}$ (which has the same poles as f) to get:

$$N_f(r, e^{i\theta}) + \log |f(0) - e^{i\theta}| = \int_0^{2\pi} \log |f(re^{i\varphi}) - e^{i\theta}| \frac{d\varphi}{2\pi} + N_f(r, \infty).$$

We integrate each side with respect to θ, and use Lemma 1.9 to find

$$\int_0^{2\pi} N_f(r, e^{i\theta}) \frac{d\theta}{2\pi} + \log^+|f(0)| = \int_0^{2\pi} \log^+|f(re^{i\varphi})| \frac{d\varphi}{2\pi} + N_f(r, \infty)$$

$$= T_f(r, \infty) \quad \text{by definition.}$$

This proves the theorem in this case. If f has a pole at 0, then the same argument yields $\log |c_f|$ instead of $\log^+|f(0)|$, as desired.

Remark. That T_f is increasing is not trivial, because the corresponding statement for m_f is not true, although it is true for N_f.

We conclude this section by showing how the classical bound with the sup norm for holomorphic functions is bounded in terms of T_f.

Let f be holomorphic on $\bar{\mathbf{D}}_R$. Let

$$M_f(R) = \log \|f\|_R, \quad \text{where} \quad \|f\|_R = \sup_{|z|=R} |f(z)|.$$

It is obvious that $T_f \leq \max(M_f, 0)$. Conversely:

Lemma 2.4. *For $r < R$ we have*

$$M_f(r) \leq \frac{R+r}{R-r} m_f(R, \infty) - \frac{R-r}{R+r} m_f(R, 0) \leq \frac{R+r}{R-r} m_f(R).$$

Proof. In Jensen's formula, since f has no poles, we can omit the terms with $\operatorname{ord}_a f < 0$ when a is a pole, so for $z = re^{i\varphi}$ and $\zeta = Re^{i\theta}$,

$$\log |f(z)| \leq \int_0^{2\pi} \log |f(Re^{i\theta})| \operatorname{Re} \frac{\zeta + z}{\zeta - z} \frac{d\theta}{2\pi}.$$

But

$$\frac{R-r}{R+r} \leq \operatorname{Re}\left(\frac{\zeta+z}{\zeta-z}\right) \leq \frac{R+r}{R-r},$$

so the lemma follows from the relation $\log \alpha = \log^+ \alpha - \log^+ 1/\alpha$, taking $\alpha = |f(Re^{i\theta})|$.

Theorem 2.5. *For f holomorphic on $\bar{\mathbf{D}}_{2r}$ we have*

$$M_f(r) \leq 3m_f(2r, \infty) = 3T_f(2r, \infty).$$

Proof. Take $R = 2r$ in the previous inequality. Since f has no poles, $N_f(r, \infty) = 0$ so $m_f = T_f$.

Theorem 2.6. *Let f be meromorphic on $\mathbf{C}$. If $T_f(R)$ is bounded, then f is constant. Also, f is a rational function if and only if*

$$T_f(R) = O(\log R) \quad \text{for} \quad R \to \infty,$$

or equivalently, there exists a sequence $R_j \to \infty$ such that

$$T_f(R_j) = O(\log R_j) \qquad for \quad j \to \infty.$$

Proof. If T_f is bounded, then from the definition $N_f \leq T_f$ is also bounded and f has no poles. Then M_f is bounded, so f is constant by Liouville's theorem. Next suppose f is rational. If $a \in \mathbf{C}$ or $a = \infty$ then $n(t, a) \leq k$, where k is the maximum of the degrees of the numerator and denominator of f. By Theorem 2.3 we get the desired order of growth for $T_f(R)$.

Conversely, suppose $T_f(R) = O(\log R)$ for $R \to \infty$, and let k be a positive integer such that

$$T_f(R) \leq k \log R + O(1).$$

Since $m_f \geq 0$, we also obtain

$$N_f(R, \infty) \leq k \log R + O(1).$$

The definition of N_f as an integral plus the term $n_f(0, \infty) \log R$ then shows that

$$n_f(t, \infty) \leq k \quad \text{for all } t.$$

In particular, f has only a finite number of poles in $\mathbf{C}$. There is a polynomial P such that Pf is entire, and we also have

$$T_{Pf}(R) = O(\log R).$$

So without loss of generality we can assume f entire. By Theorem 2.5 we conclude that $M_f(R) = O(\log R)$ also, so there is a positive integer k such that $M_f(R) \leq \log(R^k)$ for R large. By Cauchy's theorem, if $f = \sum a_n z^n$ then $|a_n| \ll \|f\|_R / R^n$, so $a_n = 0$ for $n > k$, whence f is a polynomial of degree $\leq k$, thus proving the theorem with the condition $T_f(R) = O(\log R)$. But the hypothesis $T_f(R_j) = O(\log R_j)$ can be used instead with exactly the same argument, because that is the only condition which was used above.

VI, §3. THE THEOREM ON THE LOGARITHMIC DERIVATIVE

This section contains the deepest estimate to be used, and is one of Nevanlinna's key contributions.

Theorem 3.1 (Theorem on the logarithmic derivative). *Let f be mero-morphic on* $\mathbf{C}$, *and non-constant. Then for all r outside a set of finite Lebesgue measure we have*

$$m_{f'/f}(r) \ll \log r + \log T_f(r) + \log^+ |1/c_f|,$$

where the constants implied by the sign $\ll$ are absolute. If f is mero-morphic on $\mathbf{C}$ *and of finite order, that is $T_f(r) = O(r^k)$ for $r \to \infty$ and some integer k, then*

$$m_{f'/f}(r) = O(\log r) \qquad \text{for} \quad r \to \infty$$

without any restriction.

Actually, by keeping track of the estimates of the proof, one finds that the constants are quite small, and some choice for them will be given in Proposition 3.6 below. In the applications of the next chapter, we are interested in measuring the order of growth for $r \to \infty$ when f is a given meromorphic function on $\mathbf{C}$, and one can drop the term $\log^+ |1/c_f|$ to have the inequality

$$m_{f'/f}(r) = O_f(\log r + \log T_f(r)) \qquad \text{for} \quad r \to \infty,$$

where the constant in O_f depends on f.

In Chapter VIII, we shall return to similar questions for holomorphic or meromorphic maps *on the unit disc*, where the asymptotic behavior occurs for $r \to 1$.

The rest of this section contains the proof of Theorem 3.1.

Letting $h = f/c_f$ and $T_h \leq T_f + \log^+ |1/c_f|$ we can reduce the proof to the case when $c_f = 1$, for simplicity. We then have $T_f = T_{1/f}$ by Proposition 2.1.

In what follows, we let f or h be meromorphic on the disc $\bar{\mathbf{D}}_R$ with R fixed until otherwise specified. We let

$$r < s < R,$$

and we let $|z| = r$.

We shall study systematically the logarithmic derivative f'/f when f has no zeros and poles, and when f is a canonical product.

Lemma 3.2. *Suppose h is holomorphic without zeros on $\mathbf{D}_s$. Then*

$$m_{h'/h}(r) \leq \log^+ s + 2 \log^+ \frac{1}{s-r} + \log^+ \max[m_h(s), m_{1/h}(s)] + 2 \log 2.$$

Proof. Start with the non-real formulation of Jensen's formula, which yields

$$\log h(z) = \int_0^{2\pi} \log|h(se^{i\theta})| \frac{se^{i\theta} + z}{se^{i\theta} - z} \frac{d\theta}{2\pi} + iC.$$

Differentiate with respect to z under the integral to get

$$h'/h(z) = \int_0^{2\pi} \log|h(se^{i\theta})| \frac{2se^{i\theta}}{(se^{i\theta} - z)^2} \frac{d\theta}{2\pi}.$$

Use $|\log \alpha| = \log^+ \alpha + \log^+ 1/\alpha$. Then

$$|h'/h(z)| \leq \frac{2s}{(s-r)^2} [m_h(s) + m_{1/h}(s)].$$

Now take $\log^+|h'/h(re^{i\theta})|$ and integrate to get $m_{h'/h}(r)$ on the left. From the basic inequalities of $\log^+$ of a product and a sum, we obtain the upper bound given in the right-hand side of the lemma.

Remark. The above lemma in the simple case of a function without zeros or poles will already be useful for some applications at the beginning of the next chapter. The rest of this section is not needed for these applications, except for Lemma 3.7 below.

Next we deal with the **canonical products**. Recall that

$$G_s^0(z) = \prod_{f(a) = 0} \frac{s^2 - \bar{a}z}{s(z - a)} \quad \text{and} \quad G_s^\infty(z) = \prod_{f(a) = \infty} \frac{s^2 - \bar{a}z}{s(z - a)}.$$

The products are taken with the multiplicities attached to each a and $|a| < s$. Of course we should write $G_{s,f}^0$ and $G_{s,f}^\infty$, but f will remain fixed. Also we fix s, and will omit s as a subscript. We let

$$P = P_s = G_s^\infty/G_s^0 \quad \text{and} \quad G = G_s = G_s^0 G_s^\infty.$$

Then

$$h = fP^{-1} \quad \text{has no zeros and poles in } \mathbf{D}_s \quad \text{and} \quad f = hP.$$

Since the map $f \mapsto f'/f$ is a homomorphism, we shall be able to estimate $m_{f'/f}$ in terms of $m_{h'/h}$ and $m_{P'/P}$ or $m_{G'/G}$.

Recall that each term in the product for G^0 or G^∞ has absolute value ≥ 1, so

$$\log |G^0(z)| = \log^+ |G^0(z)| \quad \text{for} \quad |z| < s,$$

and similarly for G^∞. Hence $|P| \leq |G|$ and $|1/P| \leq |G|$. We then obtain for $r \geq 1$ or all $r \geq 0$ if $f(0) \neq 0, \infty$:

$$m_h \leq m_f + m_{1/P} \leq m_f + m_G \leq T_f + m_G,$$

$$m_{1/h} \leq m_{1/f} + m_P \leq m_{1/f} + m_G \leq T_f + m_G,$$

since we assumed $c_f = 1$ so $T_{1/f} = T_f$ by Proposition 2.2. Our restriction on r is to insure that $N_f(r) \geq 0$ so $m_f \leq T_f$. Since

$$m_G(s) = 0$$

we can rewrite the inequality of Lemma 3.2 in the form of

Lemma 3.3. *Assume* $c_f = 1$. *Factor* $f = hP_s$ *where* h *has no zero or pole in* $\mathbf{D}_s$. *Then for* $r \geq 1$, *or* $r \geq 0$ *if* $f(0) \neq 0, \infty$:

$$m_{h'/h}(r) \leq \log^+ R + 2\log^+ \frac{1}{s-r} + \log^+ T_f(R) + 2\log 2.$$

This merely uses the fact that $\log^+$ and T_f are monotone increasing, and the basic inequality for $\log^+$ of a sum.

Next we give a bound for $m_{G'/G}$ and $m_{P'/P}$.

Lemma 3.4. *Let* $G = G_{s,f}$ *and* $P = P_{s,f}$. *Let*

$$n_f(s, 0 + \infty) = n_f(s, 0) + n_f(s, \infty).$$

Then for $0 < r < s$ *we have*

$m_{P'/P}(r)$ *and* $m_{G'/G}(r)$

$$\leq \log^+ \frac{s}{(s-r)^2} + \frac{s-r}{R-s} \frac{R}{r} N_f(R, 0 + \infty) + \log^+\left[\frac{R}{R-s} N_f(R, 0 + \infty)\right].$$

Proof. Consider one multiplicative term

$$G_a(z) = \frac{s^2 - \bar{a}z}{s(z-a)}.$$

Then

$$-G_a'/G_a(z) = \frac{s^2 - |a|^2}{(z-a)(s^2 - \bar{a}z)}.$$

But

$$|s^2 - \bar{a}z| \geq s^2 - |\bar{a}|r \geq s(s - r).$$

Therefore we get immediately

$$|G_a'/G_a(z)| \leq \frac{s}{(s-r)^2} |G_a(z)|.$$

We have such an inequality for each value a of a zero or a pole. We use the fact that $|G_a(z)| \geq 1$, and the fact that $Q \mapsto Q'/Q$ is a homomorphism. Then

$$|G'/G| \quad \text{and} \quad |P'/P| \leq \sum_a |G_a'/G_a| \leq \frac{s}{(s-r)^2} \sum_a |G_a|.$$

The sum is over zeros and poles. Apply $\log^+$, which is $\log$ on each $|G_a|$. Use the inequality for $\log^+$ of a sum in terms of the sum of the $\log^+$, plus the number of terms. Then integrate. We get:

$$m_{P'/P} \text{ and } m_{G'/G}(r) \leq \log^+ \frac{s}{(s-r)^2} + \sum_a m_{G_a}(r) + \log^+ n_f(s, 0 + \infty)$$

$$\leq \log^+ \frac{s}{(s-r)^2} + N_f(s, 0 + \infty) - N_f(r, 0 + \infty)$$

$$+ \log^+ n_f(s, 0 + \infty)$$

by Proposition 1.11. We now have to estimate N_f and n_f.

We formulate a general lemma which will conclude the proof of Lemma 3.4.

Lemma 3.5. *Let $n(r) \geq 0$ be a monotone increasing function of r for $0 \leq r \leq R$. For $r > 0$ let*

$$N(r) = \int_0^r [n(t) - n(0)] \frac{dt}{t} + n(0) \log r.$$

Let $0 < s < R$. If $s \geq 1$ or $n(0) = 0$ then

$$n(s) \leq \frac{R}{R-s} (N(R) - N(s)) \leq \frac{R}{R-s} N(R).$$

Similarly, for $r < s$,

$$N(s) - N(r) \leq n(s) \frac{s-r}{r} \leq \frac{s-r}{R-s} \frac{R}{r} N(R).$$

Proof. We shall use

$$\frac{R-s}{R} \leq \log \frac{R}{s} \leq \frac{R-s}{s}.$$

Then by definition

$$n(s) = \frac{1}{\log(R/s)} \, n(s) \int_s^R \frac{dt}{t} \leq \frac{1}{\log(R/s)} \int_s^R \frac{n(t)}{t} \, dt$$

$$\leq \frac{1}{\log(R/s)} \left(N(R) - N(s) \right)$$

and the first inequality follows. As to the second, for $r < s < R$:

$$N(s) - N(r) = \int_r^s n(t) \, \frac{dt}{t} \leq n(s) \log(s/r)$$

$$\leq n(s) \, \frac{s-r}{r},$$

which concludes the proof.

In the applications, $N = N_f$ and we use the inequality $N_f \leq T_f$. We now put the lemmas together. We start with

$$m_{f'/f} \leq m_{h'/h} + m_{P'/P} + \log 2,$$

and obtain

$$m_{f'/f}(r) \leq 3 \log^+ R + 4 \log^+ \frac{1}{s-r} + \log^+ \frac{1}{R-s} + 3 \log^+ T_f(R) + 4 \log 2$$

$$+ \frac{s-r}{R-s} \frac{R}{r} N_f(R, 0 + \infty).$$

The last term is obviously the worst, so we make it small and fix it up so that the other terms will be found as desired. Namely given $r < R$ we choose s such that

$$\frac{s-r}{R-s} \frac{R}{r} = \frac{\frac{1}{2}}{T+1}, \qquad \text{where} \quad T = T_f(R).$$

Assume $r \geq 1$, or $r > 0$ if $f(0) \neq 0, \infty$ to insure that $T \geq 0$. Then the last term satisfies

$$\frac{s-r}{R-s} \frac{R}{r} N_f(r, 0 + \infty) \leq 1.$$

From our choice of s, it then follows at once that s is to the left of the midpoint between r and R:

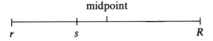

Therefore

$$\frac{1}{s-r} = \frac{R}{r}\frac{2(T+1)}{R-s} \leqq \frac{R}{r}\frac{4}{R-r}(T+1) \qquad \text{and} \qquad \frac{1}{R-s} \leqq \frac{2}{R-r}.$$

Thus we have rid ourselves of s, and get an estimate entirely in terms of r and R, namely:

Proposition 3.6. *Let f be meromorphic on $\bar{\mathbf{D}}_R$ and $c_f = 1$. Let $r \geqq 1$ or $r > 0$ if $f(0) \neq 0, \infty$. Then*

$$m_{f'/f}(r) \leqq 3\log^+ R + 4\log^+ \frac{R}{r} + 5\log^+ \frac{1}{R-r}$$

$$+ 7\log^+ T_f(R) + 17\log 2 + 1.$$

We now consider the Theorem on the Logarithmic Derivative in light of this last proposition, and let f be meromorphic on $\mathbf{C}$.

If f is of finite order, that is for some integer k

$$T_f(R) = O(R^k) \qquad \text{as} \quad R \to \infty,$$

then we can take $R = 2r$ and we obtain immediately

$$m_{f'/f}(r) = O(\log r) \qquad \text{for} \quad r \to \infty.$$

If $T_f(R)$ grows too fast, then we cannot argue this way, and in order to replace R by r we have to use the following lemma of E. Borel, which allows us to take R slightly bigger than r.

Lemma 3.7. *Let $S(r) \geqq 0$ be a continuous, non-constant, increasing function of r, for $r > 0$. Then*

$$S\left(r + \frac{1}{S(r)}\right) < 2S(r)$$

for all $r > 0$ except for r lying in a set of finite Lebesgue measure on $\mathbf{R}$.

Proof. Let E be the exceptional set where the stated inequality is false, that is $S(r + 1/S(r)) \geq 2S(r)$. Suppose there is some $r_1 \in E$, $S(r_1) \neq 0$. Let

$$r_2 = \inf\left\{r \in E \text{ such that } r \geq r_1 + \frac{1}{S(r_1)}\right\}.$$

Then $r_2 \in E$ and $r_2 \geq r_1 + 1/S(r_1)$. Let

$$r_3 = \inf\left\{r \in E \text{ such that } r \geq r_2 + \frac{1}{S(r_2)}\right\}.$$

Then $r_3 \in E$ and $r_3 \geq r_2 + 1/S(r_2)$. We continue in this way to get the sequence $\{r_n\}$.

Since S is monotone, by construction we find

$$S(r_{n+1}) \geq S\left(r_n + \frac{1}{S(r_n)}\right) \geq 2S(r_n) \geq 2^n S(r_1).$$

Hence

$$\sum \frac{1}{S(r_n)} \leq \frac{2}{S(r_1)}.$$

If the sequence $\{r_n\}$ is infinite, then it cannot be bounded, for otherwise one sees immediately that S would become infinite before the least upper bound, which does not happen since S is continuous. Then E is covered by the union of the intervals

$$\bigcup\left[r_n, r_n + \frac{1}{S(r_n)}\right]$$

which has measure $\leq 2/S(r_1)$, thereby proving the lemma.

We apply the lemma to the function

$$S(r) = \log^+ T_f(r)$$

and we take

$$R = r + \frac{1}{\log^+ T_f(r)}$$

in Proposition 3.6 to conclude the proof of Theorem 3.1. Of course we also obtain a version of Theorem 3.1 with explicit constants as follows.

Theorem 3.8. *Let f be meromorphic on $\mathbf{C}$. Suppose $r > 1$ and $T_f(r) > e$. Also $c_f = 1$. Then for all r outside a set of finite Lebesgue measure, we have*

$$m_{f'/f}(r) \leq 7 \log^+ T_f(r) + 4 \log r + 5 \log^+ \log^+ T_f(r) + 17 \log 2.$$

We won't need any such explicit result in the applications of the next chapter, but I thought it would be helpful to see how easy it is to get explicit low constants in the estimate.

VI, §4. THE SECOND MAIN THEOREM

By a **divisor** D on $\mathbf{C}$ we mean a formal linear combination

$$D = \sum_a e_D(a)(a), \qquad \text{with} \quad e_D(a) \in \mathbf{Z}, \quad a \in \mathbf{C}$$

such that any bounded region of $\mathbf{C}$ has only a finite number of points a with $e_D(a) \neq 0$. A divisor on $\mathbf{P}^1$ is a similar but finite sum, with possibly $a = \infty$.

We let D^0 and D^∞ be the divisors of zeros and poles of D, that is

$$D^0 = \sum e_D(a)(a) \qquad \text{such that} \quad e_D(a) \geq 0,$$

$$D^\infty = D^0 - D \qquad \text{so} \quad D = D^0 - D^\infty.$$

We define as usual

$$N(r, D) = \sum_{\substack{a \in \mathbf{D}_r \\ a \neq 0}} e_D(a) \log \left| \frac{r}{a} \right| + e_D(0) \log r.$$

This is the appropriately weighted r-**degree.**

Let f be a meromorphic function on $\mathbf{C}$. Then we have the **divisor of** f

$$D_f = \sum (\text{ord}_a f)(a),$$

and the divisors D_f^0 and D_f^∞ of zeros and poles of f.

Let $a \in \mathbf{C}$. Suppose f has one of the expansions:

$$f(z) = c_0 + c(z - a)^e + \text{higher terms}, \qquad \text{with } e \geq 1 \text{ and } c \neq 0;$$

$$f(z) = \frac{c}{(z - a)^e} + \text{higher terms}, \qquad \text{with } e \geq 1 \text{ and } c \neq 0.$$

Then we let $e = e(a)$ be the **multiplicity**, and $e(a) - 1$ the **ramification index** of f at a. We define the **ramification divisor** R_f to be

$$R_f = \sum (e_f(a) - 1)(a),$$

where the sum is taken over those a such that $e(a) \geq 2$. In particular, R_f is an effective divisor, that is all its coefficients are ≥ 0. We use the notation

$$N_{1,f}(r) = N(r, R_f) = \sum_{\substack{a \in \mathbf{D}_r \\ a \neq 0}} (e_f(a) - 1) \log \left| \frac{r}{a} \right| + (e_f(0) - 1) \log r,$$

and the last term with 0 occurs if and only if $e_f(0) \geq 2$. We see that $N(r, R_f)$ counts the zeros of the derivative f', plus the ramification order of f at poles. We can also write

$$R_f = D_{f'} + 2D_f^\infty.$$

The constant 2 reflects the fact that the divisor of differential forms on $\mathbf{P}^1$ has degree -2. (Cf. Chapter VII, Conjecture 5.2 and Theorem 6.1) In light of this, we have the relation

$$N_{1,f}(r) = N(r, R_f) = N_{f'}(r, 0) + 2N_f(r, \infty) - N_{f'}(r, \infty)$$

so as functions of r,

$$\boxed{N(R_f) = N_{1,f} = N_{1/f'} + 2N_f - N_{f'}.}$$

It is convenient to define

$$m_f(r, a) = \int_0^{2\pi} \log^+ \frac{1}{|f(re^{i\theta}) - a|} \frac{d\theta}{2\pi}$$

just as we had

$$m_f(r, \infty) = \int_0^{2\pi} \log^+ |f(re^{i\theta})| \frac{d\theta}{2\pi}.$$

Let

$$D = \sum_{k=1}^p (a_k)$$

be a divisor with a finite number of points with multiplicity 1. We take $a_k \in \mathbf{P}^1$, so a_k is either a complex number or ∞. We define $m_f(r, D)$ by linearity, that is

$$m_f(r, D) = \sum m_f(r, a_k).$$

A divisor D as above, where all the non-zero multiplicities are equal to 1, will be called a **simple divisor**. This is the one-dimensional version of divisors with simple normal crossings in higher dimensions.

Theorem 4.1 (Second Main Theorem). *Let f be meromorphic on $\mathbf{C}$. Let D be a simple divisor on $\mathbf{P}^1$. Then*

$$m_f(r, D) + N(r, R_f) - 2T_f(r) \leqq O_{\text{exc}}(\log r + \log T_f(r))$$

for $r \to \infty$. The O_{exc} means that the estimate holds except for r in a set of finite Lebesgue measure. If $T_f(r)$ is of finite order, then the inequality holds without any such restriction, asymptotically. The constants implicit in the O_{exc} term depend on f and D.

A much more general theorem will be proved in Chapter VII. We shall show here how Nevanlinna's theorem fits into this more general context. For comparison, we shall also give Nevanlinna's proof.

By the Weierstrass factorization theorem, we may write

$$f = f_1/f_0,$$

where f_0, f_1 are entire functions without common zeros. We then view

$$F = (f_0, f_1) : \mathbf{C} \to \mathbf{P}^1$$

as a map into the projective line. The functions f_0, f_1 are uniquely determined up to a common factor h which is an entire function without zeros.

A point of $\mathbf{C}$ or ∞ may be viewed as a hyperplane in $\mathbf{P}^1$.

Let

$$W(F) = W = f_0 f_1' - f_0' f_1$$

be the Wronskian. Then the divisor of zeros of W is precisely R_f, as may be verified at once by using the hypothesis that f_0, f_1 are entire without common zeros. Therefore

$$N(r, R_f) = N_{1/W}(r).$$

With this interpretation of the ramification term, the formula in the theorem is the special case for $\mathbf{P}^1$ of the more general formula to be proved later for $\mathbf{P}^n$.

However, as we promised, we give Nevanlinna's proof of the theorem in the present one-dimensional case in the framework of classical analysis of one variable.

Suppose first that $D = \sum (a_k)$ where $a_1, \ldots, a_p$ are complex numbers. Define

$$g(z) = \sum_{k=1}^{p} \frac{1}{f(z) - a_k}.$$

We prove a lemma:

Lemma 4.2. $m_f(r, D) \leq m_g(r) + O(1)$

Proof. Let $d = \min |a_i - a_j|$ for $i \neq j$. We first prove:

if $|f(z) - a_i| < d/4p$ then $\log^+ \dfrac{1}{|f(z) - a_i|} \leq \log^+ |g(z)| + O(1)$.

Indeed, we have

$$(f - a_i)g = 1 + \sum_{j \neq i} \frac{f - a_i}{f - a_j},$$

whence

$$\frac{1}{f - a_i} = g \left(1 + \sum_{j \neq i} \frac{f - a_i}{f - a_j} \right)^{-1}.$$

Our assertion follows from the property $m_{gh} \leq m_g + m_h$ provided we show that the sum over $j \neq i$ is small in absolute value. But

$$|f - a_j| \geq |a_i - a_j| - |f - a_i| \geq d - \frac{d}{4p} \geq \frac{3d}{4}.$$

Then

$$\left| \frac{f - a_i}{f - a_j} \right| \leq \frac{d/2p}{3d/4} \leq \frac{2}{3p}.$$

so

$$\left(1 + \sum_{j \neq i} \frac{f - a_i}{f - a_j} \right)^{-1} \leq 3.$$

This proves our assertion with $O(1) = \log 3$.

Let A_i be the arc of those θ such that $|f(re^\theta) - a_i| < d/4p$. Then

$$\sum m_f(r, a_i) \leq \sum \int_{A_i} \log^+ \frac{1}{|f - a_i|} \frac{d\theta}{2\pi} + p \log \frac{4p}{d},$$

and since the arcs A_i are disjoint $(i = 1,\ldots,p)$ the sum on the right-hand side is bounded by $m_g(r) + O(1)$, which concludes the proof of the lemma.

We come to the main part of the proof of Theorem 4.1. Suppose D consists of finite points with multiplicity 1, and consider $D + (\infty)$. Then by Lemma 4.2 we know that

$$m_f(r, D) + m_f(r, \infty) \leq m_g(r) + m_f(r) + O(1).$$

Let

$$F = \prod (f - a_k).$$

Then $m_{F'/F}(r)$ can be absorbed in the error term by the theorem on the logarithmic derivative, together with the fact that

$$T_F \leq \sum T_{f-a_k} \leq pT_f + O(1).$$

Since $F'/F = gf'$, we get

$$m_g + m_f \leq m_{gf'} + m_{1/f'} + m_f = m_{F'/F} + m_{1/f'} + m_f.$$

Therefore we conclude the proof of Theorem 4.1 with the next lemma.

Lemma 4.3.

$$m_{1/f'}(r) + m_f(r) \leq 2T_f(r) - N_{1,f}(r) + O_{f,\mathrm{exc}}(\log r + \log T_f(r)),$$

where $O_{f,\mathrm{exc}}$ has the usual meaning of O, but restricted to r outside a set of finite Lebesgue measure.

Proof. We use the lemma on logarithmic derivative to absorb all expressions of the form $m_{h'/h}$ into the error term. All subsequent inequalities are to be read modulo the error term $O_{f,\mathrm{exc}}$. We have:

$$
\begin{aligned}
m_{1/f'} + m_f &= m_{1/f'} + m_f + N_{1,f} - N_{1,f} \\
&= m_{1/f'} + m_f + 2N_f + N_{1/f'} - N_{f'} - N_{1,f} \\
&= T_{1/f'} + T_f + N_f - N_{f'} - N_{1,f} \\
&= T_{f'} - N_{f'} + T_f + N_f - N_{1,f} \quad \text{(by Proposition 2.1)} \\
&= m_{f'} + T_f + N_f - N_{1,f} \\
&\leq m_{f'/f} + m_f + T_f + N_f - N_{1,f}.
\end{aligned}
$$

We can now kill $m_{f'/f}$ and combine $m_f + N_f = T_f$ to conclude the proof.

CHAPTER VII

Applications to Holomorphic Curves in $\mathbf{P}^n$

In this chapter we start with Borel's theorem of 1897, concerning linear relations between entire functions without zeros. Its proof depends only on a very easy and brief application of Jensen's formula via Lemmas 3.2 and 3.7, and could consequently be done in standard basic courses in complex variables.

We then give several applications, first in the context of holomorphic curves $f: \mathbf{C} \to \mathbf{P}^n$ whose images miss hyperplanes. Some of these results have their roots in Bloch [Bl] and Cartan [Ca 1], which were extended by Fujimoto [Fu] and Green [Gr 1], [Gr 2]. In general, we give proofs of results which had been stated in Chapter II, §1 and Chapter III, §3, to provide examples of the general theorems of those chapters. We also give the example of the Fermat hypersurface, including Green's theorem that if the degree is d in $\mathbf{P}^n$ and $d \geq n^2$ then the only holomorphic maps of $\mathbf{C}$ into Fermat are "trivial" ones. In the same vein, we prove that the Brody–Green perturbation of the Fermat hypersurface is hyperbolic.

Although the First Main Theorem for holomorphic curves $\mathbf{C} \to X$ into arbitrary non-singular varieties is essentially an immediate consequence of the case when $X = \mathbf{P}^1$, the Second Main Theorem is known today only in very special cases. We prove one significant case due to Cartan, relating to hyperplanes in general position. I find it very unfortunate that to a large extent Cartan's paper [Ca 2] has been overlooked by many people for many years, since the Ahlfors and Weyl proofs which came much later are also much more complicated. Sections 3, 5, and 6 form a self-contained whole and could be read immediately without referring to the other sections which deal with concrete cases. Many readers might prefer such an arrangement, which gives a more rapid structural insight into what is going on.

The theory and proofs of this chapter are linear. The main technique of proof using Wronskians is due to Nevanlinna [Ne 2] and Bloch [Bl]. Further results of Bloch, considerably expanded by Cartan, will be treated in the next chapter.

Vojta has pointed out the striking similarity between the Second Main Theorem in the context of value distribution of holomorphic curves, and the theory of heights in number theory. In the exposition of the notions in this last part of the chapter, I use a language and notation parallel to that used in this theory of heights. The CR note where Cartan announced his Second Main Theorem for the case of hyperplanes was published in 1929, essentially at the same time as Weil's thesis in 1928, where he uses the height which today bears his name and which was defined simultaneously by Siegel [Si]. But no one at the time saw that Cartan's definition of this height was entirely analogous to the definition of the heights in algebraic number theory, and that both were based on the product formula. This gap of understanding is, to me, almost as striking as the gap of understanding between Artin and Hecke in Hamburg about the connection between non-abelian L-series and modular forms. One had to await 40 to 50 years for the connections to be made, conjecturally, by Langlands and Vojta respectively in these two cases. In both cases, some algebraic number theorist's failure to relate properly to analysis (and conversely) contributed to that gap in understanding.

VII, §1. BOREL'S THEOREM

Let $h_0, \ldots, h_n$ be entire functions without zeros, which we call **units**, because they are units in the ring of entire functions. We consider the possibility that they satisfy the equation

$$h_0 + \cdots + h_n = 0.$$

First we describe trivial solutions of this equation. Start with one unit h, and let $c_0 = 1, \ldots, c_s$ be non-zero constants such that

$$\sum_{i=0}^{s} c_i = 0.$$

Let $h_i = c_i h$. Then $h_0 + \cdots + h_s = 0$. By taking a partition of $\{0, \ldots, n\}$ into subsets S such that for each S we are in the above situation we obtain what we call **trivial solutions** of the equation in units. Borel's theorem states that these are the only solutions.

Theorem 1.1. *Let $h_0, \ldots, h_n$ be units. Suppose*

$$h_0 + \cdots + h_n = 0.$$

Define an equivalence $i \sim j$ if there exists a constant c (necessarily $\neq 0$) such that $h_i = ch_j$. Let $\{S\}$ be the partition of $\{0, \ldots, n\}$ into equivalence classes. Then

$$\sum_{i \in S} h_i = 0.$$

If $n \leq 2$ then there is only one equivalence class.

Borel's theorem is a special case of theorems which will be proved later, and could be deduced as a corollary. For instance, let d be a large positive integer compared to n. A unit h can be written as the d-th power of a unit, because $\log h$ is defined as an entire function, so $h = g^d$ where $g = \exp((\log h)/d)$. This reduces Borel's theorem to the analogous result for the Fermat equation with arbitrarily large d, to be handled in §4. Borel's theorem will also be deduced as a corollary of Cartan's theorem in §6.

This trick of taking d-th power, to reduce a linear equation to one of higher degree seems to have been first used by Siegel [Si] implicitly, in the context of number theory. I believe it was first brought out explicitly in terms of the "unit equation" in [La 0], and *Diophantine Geometry*.

In writing this chapter, I had to make a decision whether to state and prove the Second Main Theorem of §6 first, and then obtain all the other results in this framework, or essentially develop first the unit equation and the Fermat equation, building up to the Second Main Theorem. I finally chose this latter development.

For the proof we shall need two corollaries of the Theorem on Logarithmic Derivative. We know that for any meromorphic function f we have

$$m_{f'/f}(r) = O_{\text{exc}}(\log r + \log^+ T_f(r))$$

for $r \to \infty$, where O_{exc} means for r outside an exceptional set E of finite measure. For simplicity of notation, we sometimes omit the subscript exc in the course of arguments, although we keep the subscript in the formal statement of results. Thus O and o may mean O_{exc} and o_{exc} in the course of a proof.

We apply the estimate when $f = h$ is a unit. Then $m_h = T_h$ since h has no poles. Thus

$$T_{h'/h}(r) = m_{h'/h}(r) = O_{\text{exc}}(\log r + \log^+ T_h(r)).$$

Given a set E of finite measure and a constant C, suppose there are arbitrarily large $r \notin E$ such that

$$T_h(r) \leq C \log r.$$

Then h is a polynomial, whence h is constant since h is a unit. Therefore, if h is not constant, given C we have

$$T_h(r) \geq C \log r \qquad \text{for all} \quad r \geq r_0(C), \quad r \notin E.$$

This implies that

$$T_{h'/h}(r) = o(T_h(r)) \qquad \text{for} \quad r \notin E.$$

We state this as a separate result.

Proposition 1.2. *Let h be a unit. If h is not constant, then*

$$T_{h'/h}(r) = o_{\text{exc}}(T_h(r)).$$

In addition, we shall need a corollary of Nevanlinna estimates.

Proposition 1.3. *Let g be meromorphic on* $\mathbf{C}$, $g \neq 0$. *Then*

$$T_{g'} = O_{\text{exc}}(T_g).$$

Proof. By Theorem 3.1 of Chapter VI, we have

$$m_{g'/g}(r) = O(\log r + \log^+ T_g(r)).$$

If there exists a constant C such that $T_g(r) \leq C \log r$ for arbitrarily large r outside the exceptional set, then g is a rational function, and $m_{g'/g} = O(T_g)$. One verifies at once directly that $N_{g'} = O(N_g)$, and the desired estimate follows. If there does not exist a constant C as above, then

$$\log r = o(T_g(r)),$$

whence

$$m_{g'/g} = o(T_g).$$

Since again $N_{g'} = O(N_g) = O(T_g)$ the proposition follows.

We shall now give one proof of Theorem 1.1 by the Wronskian technique of Nevanlinna [Ne 2], pursued by Bloch [Bl]. This technique will be used again in §3, §4, and §6. We begin by general remarks concerning the Borel equation but with arbitrary meromorphic functions.

Let $f_1, \ldots, f_n$ be meromorphic functions (but to a large extent what we do depends only on formal properties of derivations). We recall that their **Wronskian** is

$$W(f) = W(f_1, \ldots, f_n) = \begin{vmatrix} f_1 & \cdots & f_n \\ f_1' & \cdots & f_n' \\ \vdots & & \vdots \\ f_1^{(n-1)} & \cdots & f_n^{(n-1)} \end{vmatrix}.$$

Suppose that $f_i \neq 0$ for all i, and that

$$f_1 + \cdots + f_n = 1.$$

Taking $n - 1$ derivatives yields a system of linear equations:

$$f_1 + \cdots + f_n = 1$$
$$(f_1'/f_1)f_1 + \cdots + (f_n'/f_n)f_n = 0,$$
$$\cdots$$
$$(f_1^{(n-1)}/f_1)f_1 + \cdots + (f_n^{(n-1)}/f_n)f_n = 0.$$

Let

$$L(f_1, \ldots, f_n) = \begin{vmatrix} 1 & \cdots & 1 \\ f_1'/f_1 & \cdots & f_n'/f_n \\ \vdots & & \vdots \\ f_1^{(n-1)}/f_1 & \cdots & f_n^{(n-1)}/f_n \end{vmatrix}$$

and let

$$L_i(f_1, \ldots, f_n) = \begin{vmatrix} 1 & \cdots & 1 & \cdots & 1 \\ f_1'/f_1 & \cdots & 0 & \cdots & f_n'/f_n \\ \vdots & & \vdots & & \vdots \\ f_1^{(n-1)}/f_1 & \cdots & 0 & \cdots & f_n^{(n-1)}/f_n \end{vmatrix}.$$

Then $L(f)$ is the usual Wronskian divided by the product $f_1 \cdots f_n$, that is

$$L(f) = W(f)/f_1 \cdots f_n,$$

and similarly for $L_i(f)$.

If $f_1, \ldots, f_n$ are linearly independent, then $W(f) \neq 0$ and so $L(f) \neq 0$. We have

$$f_i = L_i(f)/L(f).$$

The L is to suggest logarithmic derivative.

In the next lemma, we first deal with the determinants $L = L(f)$ and $L_i = L_i(f)$ which are defined without any particular assumption on f itself.

Lemma 1.4. *Let $f_1, \ldots, f_n$ be meromorphic functions. Let $L = L(f)$ and $L_i = L_i(f)$ be the above determinants. Let $T = T_{f_1} + \cdots + T_{f_n}$. Then*

$$m_L(r) \text{ and } m_{L_i}(r) = O_{\text{exc}}(\log r + \log^+ T(r)).$$

Proof. Let g be a meromorphic function. We can write a logarithmic derivative of higher order as a product

$$\frac{g^{(k)}}{g} = \frac{g^{(k)}}{g^{(k-1)}} \cdots \frac{g^{(1)}}{g}.$$

By Proposition 1.3 we know that $T_{g'} = O(T_g)$ for $g = f_i, f_i', f_i^{(2)}, \ldots$. The lemma follows at once from the theorem on logarithmic derivatives.

Note. For another, equally easy and constructive argument, see Chapter VIII, §1.

Now suppose $f_i = h_i$ is a unit for each i, and suppose that

$$h_1 + \cdots + h_n = 1.$$

As we have seen in Proposition 1.2, we have

$$\log r = o(T(r)),$$

otherwise all h_i are constant and we are done. But if $h_1, \ldots, h_n$ are linearly independent, then $h_i = L_i/L$ and

$$m_{h_i} \leq m_{L_i} + m_{1/L}.$$

For $r \geq 1$ we have $N_{1/L}(r) \geq 0$ so $m_{1/L} \leq T_{1/L} = T_L + O(1)$. But since $h_1, \ldots, h_n$ are units, L has no pole, so $T_L = m_L$. Hence by Lemma 1.4 we get

$$m_{h_i}(r) = O(\log r + \log^+ T(r)),$$

whence $T_{h_i} = o(T)$ for each i and therefore $T = o(T)$, which is a contradiction. Therefore $h_1, \ldots, h_n$ are linearly dependent.

We can then proceed by induction. From a linear relation

$$c_1 h_1 + \cdots + c_n h_n = 0$$

with not all $c_i = 0$ we can deduce a relation, say

$$a_1(h_1/h_n) + \cdots + a_m(h_m/h_n) = 1,$$

with constants $a_i \neq 0$ for all i. By induction it follows that there is some constant $c \neq 0$ and $i \neq j$ such that

$$(h_i/h_n)/(h_j/h_n) = c = h_i/h_j.$$

If we now combine $h_i + h_j = (1 + c)h_j$ in the original relation

$$h_1 + \cdots + h_n = 1$$

we can continue by induction to prove Borel's theorem.

The significance of the modified Wronskians as projective coordinates lies in part in the following result.

Proposition 1.5. *Let* $f_1, \ldots, f_n$ *be meromorphic. Let*

$$L(f_1, \ldots, f_n) = W(f_1, \ldots, f_n)/f_1 \cdots f_n,$$

where W is the Wronskian. Then for any function g,

$$W(gf_1, \ldots, gf_n) = g^n W(f_1, \ldots, f_n)$$

and

$$L(gf_1, \ldots, gf_n) = L(f_1, \ldots, f_n).$$

Thus L is homogeneous of degree 0. The formula for L follows from that of W, which is immediately proved by row operations. We subtract appropriate multiples of all the rows preceding a given row, in order to get rid of the mixed terms in the expansion of $(gf)^{(k)}$ leaving only the term $gf^{(k)}$. This explains the invariant role played by the determinants L, which depend only on the point $(f_1, \ldots, f_n)$ in projective space. To see a continuation of this train of thought, the reader may skip immediately to §4 and §6.

Next in this section, we give another proof of a more general version, mostly due to Borel [Bo], who did not necessarily take constant coefficients, and did not use Wronskians, but a direct linear elimination. For the details of the proof in Nevanlinna language, and the precise statement of the theorem, I follow Mark Green [Gr 1].

For the proof, we recall that the logarithmic derivative $f \mapsto f'/f$ is a homomorphism, and if two functions have the same logarithmic derivative, then one is a constant multiple of the other.

We specify $o = o_{\text{exc}}$ in statements of propositions or lemmas, but usually omit the subscript exc in proofs for simplicity.

Theorem 1.6. *Let $h_1, \ldots, h_n$ be units. Let $g_1, \ldots, g_n$ be meromorphic functions on C. Let $n \geq 1$. Let*

$$T = \sum_{i=1}^{n} T_{h_i}.$$

Assume that $T_{g_i} = o_{\text{exc}}(T)$ for each i if T is unbounded, and otherwise T_{g_i} is bounded. Suppose we have a relation

$$g_1 h_1 + \cdots + g_n h_n = 1.$$

Then there exists a proper subset I of the indices $\{1, \ldots, n\}$ and constants $c_i \neq 0$ $(i \in I)$ such that

$$\sum_{i \in I} c_i g_i h_i = \text{constant}.$$

If $n = 1$ or 2, then h_1, h_2 are constant, and so are g_1, g_2.

Proof. Suppose first $n = 1$. Write $gh = 1$. Then

$$T_g = T_{1/h} = T_h + O(1),$$

and in light of the hypothesis we conclude that $T_h = o(T_h)$ or T_h is bounded, so h is constant.

Suppose $n \geq 2$. We differentiate the original relation to get

$$\sum_{i=1}^{n} f_i h_i = 0 \qquad \text{where} \quad f_i = g_i' + g_i h_i'/h_i.$$

If $f_i = 0$ for some i then $g_i h_i$ is constant, and we are done. Suppose $f_i \neq 0$ for all $i = 1, \ldots, n$. Then we get the relation

$$\sum_{i=1}^{n-1} (f_i/f_n) h_i/h_n = -1.$$

We have to verify the induction hypothesis for this shorter relation, with the coefficients f_i/f_n relative to the units $h_1/h_n, \ldots, h_{n-1}/h_n$, that is

$$T_{f_i/f_n} = o(T_{h_1/h_n} + \cdots + T_{h_{n-1}/h_n})$$

or $O(1)$. Let us postpone this for a moment and draw the conclusions from it.

If $n = 2$ so $n - 1 = 1$, we get

$$(f_1/f_2) = -h_2/h_1.$$

Since we are assuming $T_{f_1/f_2} = o(T_{h_2/h_1})$ or $O(1)$ it follows that h_2/h_1 is constant. Then T_{h_2} and T_{h_1} have the same order of magnitude. Hence we find

$$T_{h_1} + T_{h_2} = o(T_{h_2} + T_{h_1}) + O(1),$$

whence T_{h_i} is bounded (up to the exceptional set) for $i = 1, 2$ and hence h_i is constant for $i = 1, 2$. Finally g_i is constant for $i = 1, 2$ because $T_{g_i} = o(T_{h_i})$ or $O(1)$. This proves the case $n = 2$.

Next suppose $n \geq 3$ and T is unbounded. By induction, there is a proper subset I of $\{1, \ldots, n-1\}$ and constants $c_i \neq 0$ such that

$$\sum_{i \in I} c_i(f_i/f_n)h_i/h_n = c \quad \text{(constant)}.$$

This yields

$$\sum_{i \in I} c_i f_i h_i - c f_n h_n = 0.$$

But $f_i h_i = (g_i h_i)'$. Integrating yields the linear relation asserted in the theorem.

There remains to prove f_i/f_n is constant or

$$T_{f_i/f_n} = o(T^{(n-1)}) \qquad \text{where} \quad T^{(n-1)} = \sum_{i=1}^{n-1} T_{h_i/h_n}.$$

From the original relation $h_n^{-1} = g_n + \sum_{i=1}^{n-1} g_i h_i/h_n$ we get

$$T_{h_n} = T_{1/h_n} + O(1) \leq \sum_{i=1}^{n} T_{g_i} + \sum_{i=1}^{n-1} T_{h_i/h_n} + O(1).$$

From the standard formalism of T we know that

$$(*) \qquad T_{h_i} \leqq T_{h_i/h_n} + T_{h_n} \qquad \text{for all} \quad i = 1,\ldots,n.$$

Therefore by assumption that g_i is constant or $T_{g_i} = o(T)$ we get

$$T_{h_n} \leqq o(T_{h_n}) + o\left(\sum_{i=1}^{n-1} T_{h_i/h_n}\right) + O\left(\sum_{i=1}^{n-1} T_{h_i/h_n}\right) + O(1),$$

so

$$(**) \qquad T_{h_n} \ll \sum_{i=1}^{n-1} T_{h_i/h_n} + O(1).$$

By Propositions 1.2 and 1.3, and the fact that f_i/f_n is composed of logarithmic derivatives h_i'/h_i, h_n'/h_n and derivatives g_i', we conclude that

$$T_{f_i/f_n} = o\left(\sum_{i=1}^{n} T_{h_i}\right).$$

The desired estimate now follows from $(*)$ and $(**)$. This concludes the proof of Theorem 1.6.

Theorem 1.1 is an immediate consequence of Theorem 1.6. Indeed, under the hypothesis of Theorem 1.1, induction shows that there exists a pair of indices $i \neq j$ such that $h_i = ch_j$ for some constant c. We may thus shorten the relation, and use induction to conclude the proof.

Note that Theorem 1.1 is sometimes used in the form

$$h_1 + \cdots + h_n = 1.$$

In this case, some h_i $(i = 1,\ldots,n)$ has to be constant, and one of the equivalence classes of indices consists of those i such that h_i is constant. If we let K be this class, then we have a partition

$$\{1,\ldots,n\} = K \cup S_1 \cup \cdots \cup S_q,$$

where the equivalence classes $S_1,\ldots,S_q$ correspond to the non-constant functions. We have $q = 0$ if and only if h_i is constant for all i. In general,

$$\sum_{k \in K} h_k = 1 \qquad \text{and} \qquad \sum_{i \in S} h_i = 0 \qquad \text{if} \quad S \neq K.$$

VII, §2. HOLOMORPHIC CURVES MISSING HYPERPLANES

We shall interpret Borel's theorem in the context of holomorphic maps of $\mathbf{C}$ into $\mathbf{P}^n$, and look closer into the linear relations.

Let $(x_0,\dots,x_n)$ be the coordinates of $\mathbf{P}^n$. Let

$$f: \mathbf{C} \to \mathbf{P}^n$$

be holomorphic. Then on each affine subset $\mathbf{A}_i$ $(i = 0,\dots,n)$ defined by $x_i \neq 0$ we have a holomorphic map

$$f: U_i \to \mathbf{A}_i,$$

where U_i is the open subset of $\mathbf{C}$ where the i-th coordinate of f is non-zero. We can write f in non-homogeneous coordinates

$$f = (g_0,\dots,g_{i-1},1,g_{i+1},\dots,g_n).$$

For each $z \in \mathbf{C}$ there is an index i such that all these coordinate functions are holomorphic at z. Then any ratio g_j/g_k is meromorphic at z. Therefore the coordinate functions $g_0,\dots,g_n$ above are meromorphic functions on $\mathbf{C}$. By Weierstrass' Factorization Theorem, there exists an entire function g such that the functions $gg_0,\dots,gg_n$ are entire, and have no zeros in common. Therefore f is represented by a holomorphic map

$$F: \mathbf{C} \to \mathbf{A}^{n+1}$$

whose coordinates are $(f_0,\dots,f_n)$ with $f_i = gg_i$, and such that $f_0,\dots,f_n$ have no zeros in common. We may say that we have **lifted** f **to** $\mathbf{A}^{n+1}$.

Let H_k $(k = 0,\dots,m)$ be hyperplanes of $\mathbf{P}^n$. We recall that they are said to be in **general position** if $m \geq n$ and any $n + 1$ of these hyperplanes are linearly independent.

Suppose $m = n + 1$, so we deal with $n + 2$ hyperplanes in general position $H_0,\dots,H_{n+1}$. Then there is a projective change of coordinates such that these hyperplanes are defined by the equations

$$x_0 = 0,$$
$$\vdots$$
$$x_n = 0,$$
$$x_0 + \cdots + x_n = 0.$$

Indeed, since $H_0, \ldots, H_n$ are linearly independent, we can make a change of coordinates such that they become the coordinate hyperplanes, and then for H_{n+1} we simply make scalar dilations on the coordinates to put it in the above form, which we call **standard**.

Also we recall that given a subset of indices $J = \{j_1, \ldots, j_q\}$ with $2 \leq q \leq n$, we define the corresponding **diagonal** D_J by

$$x_{j1} + \cdots + x_{j_q} = 0.$$

We can reformulate part of Borel's theorem in the form:

Theorem 2.1 (Bloch–Cartan). *Let* $f: \mathbf{C} \to \mathbf{P}^n$ *be non-constant holomorphic with* $n \geq 2$. *Let* $H_0, \ldots, H_{n+1}$ *be* $n + 2$ *hyperplanes in general position. If the image of* f *lies in the complement of* $H_0 \cup \cdots \cup H_{n+1}$ *then it lies in some diagonal hyperplane.*

Proof. Let the hyperplanes be in standard form as above. Let f be represented by coordinates $(f_0, \ldots, f_n)$ where f_i is entire for all i and the f_i have no common zero. The assumption implies that $f_i(z) \neq 0$ for all z, and also

$$f_0(z) + \cdots + f_n(z) \neq 0 \quad \text{for all } z$$

after taking the hyperplanes normalized as above. Let

$$h_i = \frac{f_i}{f_0 + \cdots + f_n}.$$

Then

$$\sum_{i=0}^{n} h_i = 1.$$

By hypothesis, one of the partitions S of Theorem 1.1 corresponding to non-constant functions is not empty, and the linear relation

$$\sum_{i \in S} h_i = 0$$

is the defining relation of a diagonal hyperplane. This proves the theorem.

For convenience, we took f mapping $\mathbf{C}$ into $\mathbf{P}^n$ with $n \geqq 2$. But with $n = 1$ we have the stronger:

Theorem 2.2 (Picard's theorem). *Let $f \colon \mathbf{C} \to \mathbf{P}^1 - \{$three points$\}$ be holomorphic. Then f is constant.*

Proof. The proof is the same, using the corresponding case of Theorem 1.1. It was Borel who put Picard's theorem in this context.

Of course, from other theories one knows that the universal covering space of $\mathbf{P}^1 - \{$three points$\}$ is the disc, so we see Theorem 2.2 also from this other side. We are now approaching the theory of hyperbolic spaces from the Nevanlinna point of view, and we have proved:

Let X be the complement in $\mathbf{P}^n$ of $n + 2$ hyperplanes in general position. Let Y be the union of the diagonals. Then X is Brody hyperbolic modulo Y for $n \geqq 2$.

Next we give a statement which further qualifies the image of f. We shall follow Green's proof of this theorem.

Theorem 2.3 (Fujimoto [Fu 1] and Green [Gr 1]). *Let $f \colon \mathbf{C} \to \mathbf{P}^n$ be holomorphic. Assume that the image of f lies in the complement of $n + 2$ hyperplanes in general position. Then the image of f is contained in a projective linear subspace of dimension $\leqq [n/2]$. More generally, if the image of f lies in the complement of $n + p$ hyperplanes in general position, then this image is contained in a linear subspace of dimension $\leqq [n/p]$.*

Proof. Consider first the case $p = 2$, and take the hyperplanes in standard form as previously. We continue the notation of Theorem 2.1. We partition $\{0, \ldots, n\}$ into Borel equivalence classes

$$S_1 \cup \cdots \cup S_q \cup K,$$

where K is the class of the constant functions, and $i \sim j$ if h_i/h_j is constant. Given a class $S \neq K$, fix an index i in S. For $j \in S$ and $j \neq i$, we get a linear relation

$$x_j - c_j x_i = 0.$$

Let s_k be the number of elements in S_k, with $s_0 = \operatorname{card}(K)$. Then we have at least the following number of independent linear relations:

$$(s_1 - 1) + \cdots + (s_q - 1) + (s_0 - 1) = n + 1 - (q + 1)$$

$$= n - q$$

$$\geqq n/2,$$

with this last inequality being immediate because $n \geqq 2q$. This proves the first part.

Essentially the same type of argument works in general. Let $H_1(x), \ldots, H_{n+p}(x)$ be the linear forms defining the hyperplanes $H_k(x) = 0$, where the homogeneous coordinates are

$$(x) = (x_0, \ldots, x_n).$$

Any $n + 1$ of these are linearly independent, and any $n + 2$ satisfy a linear relation with coefficients which are all $\neq 0$ (otherwise $n + 1$ hyperplanes would be linearly dependent). Let $f = (f_0, \ldots, f_n)$ with $f_0, \ldots, f_n$ entire without common zeros. We then obtain $n + p$ functions

$$H_k(f) = h_k \qquad (k = 1, \ldots, n + p)$$

which are units. We partition the set of indices according to the equivalence relation $i \sim j$ if $h_i = c h_j$ for some constant c. Thus

$$\{1, \ldots, n + p\} = \bigcup S = S_1 \cup \cdots \cup S_q,$$

where S denotes an equivalence class.

We first claim that the complement of a given S has at most n elements. Otherwise, suppose there are at least $n + 1$ elements in the complement. Pick $n + 1$ such elements and one element in S to give a set of $n + 2$ indices which we denote by J, so $J \cap S$ has exactly one element. There is one relation

$$\sum_{j \in J} a_j H_j = 0, \qquad \text{with} \quad a_j \neq 0 \quad \text{for all} \quad j \in J,$$

whence a relation

$$\sum_{j \in J} a_j h_j = 0.$$

This contradicts Borel's theorem. Hence $\{1, \ldots, n + p\} - S$ has at most n elements. Hence S has at least p elements. If q is the number of equivalence classes of indices, it follows that $q \leqq (n + p)/p$.

Let T be any subset of $\{1,\ldots,n+p\}$ consisting of $n+1$ elements. Then the forms H_i $(i \in T)$ are linearly independent. Write

$$T = T_1 \cup \cdots \cup T_q \qquad \text{where} \qquad T_k = T \cap S_k.$$

Let $t_k = \text{card } T_k$. Each T_k gives rise to $t_k - 1$ equations so we obtain at least

$$t_1 - 1 + \cdots + t_q - 1 \qquad \text{linearly independent equations}$$

$$\geqq n + 1 - q \geqq n + 1 - \frac{n+p}{p} = n - \frac{n}{p}.$$

Hence the dimension of the intersection is $\leqq n/p$, thus proving the theorem.

Corollary 2.4. *Let* $f : \mathbf{C} \to \mathbf{P}^n$ *be holomorphic. If the image of* f *lies in the complement of* $2n + 1$ *hyperplanes in general position, then* f *is constant.*

Proof. Take $p = n + 1$ in the theorem.

In the language of Brody hyperbolic spaces, we can state Corollary 2.4 as follows:

Let V *be the complement in* $\mathbf{P}^n$ *of* $2n + 1$ *hyperplanes in general position. Then* V *is Brody hyperbolic.*

But we are in a position to show that the hypotheses in Green's general theorem given as Theorem 3.6 of Chapter III are satisfied, namely:

Theorem 2.5 (Bloch [Bl], Green [Gr 3]). *Let* $X_1,\ldots,X_{2n+1}$ *be* $2n + 1$ *hyperplanes in* $\mathbf{P}^n$ *in general position. Let*

$$X = X_1 \cup \cdots \cup X_{2n+1}$$

be their union. Then:

(a) $\mathbf{P}^n - X$ *is Brody hyperbolic.*
(b) *For each choice of distinct indices*

$$\{i_1,\ldots,i_k,j_1,\ldots,j_r\} = \{1,\ldots,2n+1\}$$

   the complex space $X_{i_1} \cap \cdots \cap X_{i_k} - (X_{j_1} \cup \cdots \cup X_{j_r})$ *is Brody hyperbolic.*
(c) $\mathbf{P}^n - X$ *is complete hyperbolic and hyperbolically imbedded in* $\mathbf{P}^n$.

Proof. Let $I = \{i_1, \ldots, i_k\}$ and let

$$X_I = \bigcap_{i \in I} X_i.$$

We can identify X_I with a projective space of dimension $n - k$. The intersections of the hyperplanes X_j with X_I for $j \notin I$ are hyperplanes in general position in X_I. This is immediate from the definition. Hence (a) and (b) are special cases of Corollary 2.4. Then (c) follows from Green's Theorem 3.6 of Chapter III.

We give one further application of the Borel theorem due to Green [Gr 2].

Let $X \subset \mathbf{P}^n$ be an algebraic subvariety of $\mathbf{P}^n$, by which we mean an irreducible algebraic subset. Let $P_1, \ldots, P_m$ be homogeneous polynomials determining hypersurface sections

$$Y_i = X \cap (P_i = 0)$$

of X. We say that these hypersurface sections are **non-redundant** if no one of them is contained in the union of the others.

Theorem 2.6 (Green [Gr 2]). *Let $f : \mathbf{C} \to X \subset \mathbf{P}^n$ be a holomorphic map which omits $\dim X + 2$ hypersurface sections, assumed non-redundant. Then the image of f lies in a proper algebraic subset of X.*

Proof. Let $d = \dim X$ and $P_1, \ldots, P_{d+2}$ the homogeneous polynomials defining the hypersurface sections. Raising each of these polynomials to a suitable power, we may assume that they have the same degree. Quotients P_i/P_j then define rational functions on X, and since X has dimension d, the transcendence degree of its function field, it follows that there exists a non-zero homogeneous polynomial Q in $d + 2$ variables such that

$$Q(P_1, \ldots, P_{d+2}) = 0 \quad \text{on } X.$$

Write $f = (f_0, \ldots, f_n)$ where the f_i are entire without common zeros. Then

$$P_1(f) = h_1, \ldots, P_{d+2}(f) = h_{d+2}$$

are $d + 2$ entire functions without zeros. Furthermore,

$$Q(h_1, \ldots, h_{d+2}) = 0.$$

Write Q as a sum of monomials

$$Q(t_1, \ldots, t_{d+2}) = \sum c_M M(t),$$

where $M(t)$ is a monomial

$$M(t) = t_1^{m_1} \cdots t_{d+2}^{m_{d+2}},$$

and the coefficients c_M are constant, not all 0. Each $M(h)$ is an entire function without zeros. By the Borel theorem, there exists a linear relation among two distinct monomials, that is

$$h_1^{m_1} \cdots h_{d+2}^{m_{d+2}} = c h_1^{k_1} \cdots h_{d+2}^{k_{d+2}}$$

with some constant c, with some $k_i \neq m_i$. If, say, $m_1 > k_1$ we divide both sides by $h_1^{k_1}$, and similarly for the other indices. We then end up with a relation

$$\prod_{i \in I} h_i^{r_i} = c \prod_{i \notin I} h_i^{r_i}$$

with exponents $r_i > 0$ and a suitable set of indices I. This implies that the polynomial

$$R = \prod_{i \in I} P_i^{r_i} - c \prod_{i \notin I} P_i^{r_i}$$

vanishes on the image of f. On the other hand, R does not vanish on X, for otherwise one of the hypersurface sections $P_i = 0$ of X would be redundant. This concludes the proof.

In [Gr 2] the reader will find other similar applications to maps of $\mathbf{C}$ into Grassmannians, and to Big Picard theorems. In the next two sections, we still give the applications to the Fermat case and the Brody–Green example, which are of special significance.

VII, §3. THE HEIGHT OF A MAP INTO $\mathbf{P}^n$

Before coming to the special equation we have to discuss systematically the height of a map into projective space. This might have been done much earlier, but the structure of projective space did not yet really come into play.

Let F be a field with a family of absolute values $\{v\}$ satisfying the product formula. Let $x = (x_0, \ldots, x_n)$ be a point with coordinates $x_j \in F$, not all 0. We define the **height**

$$h(x) = \sum_v \log \max_j \|x_j\|_v = \sum_v \max_j -v(x_j).$$

The product formula guarantees that this height depends only on the point in projective space $\mathbf{P}^n(F)$, since the term which will come out as a result of multiplying all coordinates by some non-zero element of F will be 0.

Example. In number theory, the fundamental example is when $F = \mathbf{Q}$ is the field of rational numbers, and the set of absolute values consists of the ordinary real absolute value ("at infinity") and all p-adic absolute values. In that case, given a point $x \in \mathbf{P}^n(\mathbf{Q})$, we can represent x with coordinates $(x_0, \ldots, x_n)$ as above such that $x_0, \ldots, x_n$ are relatively prime integers. Then the height can be expressed entirely in terms of the absolute value at infinity, namely

$$h(x) = \log \max_j |x_j|.$$

We shall carry out the same construction in the Nevanlinna case, when F is the field of meromorphic functions on $\bar{\mathbf{D}}_r$. Since the product formula is not quite satisfied because of the leading coefficient, as we saw in Chapter VI, §1, we have to introduce the corresponding perturbation in defining the height. Also, we shall follow the analysts' notation, and write T instead of h for the height.

Let

$$f : \mathbf{C} \to \mathbf{P}^n, \qquad f = (f_0, \ldots, f_n)$$

be a holomorphic map into projective space, with meromorphic coordinates $f_0, \ldots, f_n$. As we have seen, one can find an equivalent map such that all f_i are holomorphic and have no zeros in common. By

$$f = (f_0, \ldots, f_n) = (g_0, \ldots, g_n) = g$$

we mean that there exists a meromorphic function h such that $f_i = h g_i$ for all $i = 0, \ldots, n$. Thus f and g define the same map into $\mathbf{P}^n$.

In Chapter VI, we defined absolute values and valuations by:

$$\|g_i\|_{\theta, r} = |g_i(r e^{i\theta})|$$

$$-\log \|g_i\|_{a, r} = v_{a, r}(g_i) = (\operatorname{ord}_a g_i) \log \left| \frac{r}{a} \right|.$$

We now define the **Cartan height** of the map f by the formula

$$T_f(r) = \int_0^{2\pi} \log \max_i \|f_i\|_{\theta,r} \frac{d\theta}{2\pi} - \log \max_i |f_i(0)|$$

under the assumption that $f_0, \ldots, f_n$ are entire without common zero.

More generally, in terms of the meromorphic coordinates $(g_0, g_1, \ldots, g_n)$ we define this **height** by the formula

$$T_f(r) = \int_0^{2\pi} \log \max_i \|g_i\|_{\theta,r} \frac{d\theta}{2\pi} - \log \max_{i \in M} |c_{g_i}| \\ + \sum_{a \in \mathbf{D}_r} \log \max_i \|g_i\|_{a,r}$$

with the following notation. Let k be the minimal order at 0 of the functions g_i. Then we let M be the set of indices i such that g_i has order k at 0. If $g_i = f_i$ with $f_0, \ldots, f_n$ entire having no common zeros, then we get back the other expression

$$\log \max_{i \in M} |c_{g_i}| = \log \max_i |f_i(0)|.$$

Furthermore in this case the sum over $a \in \mathbf{D}_r$ is equal to 0. However, it is necessary to use the formula with the meromorphic g_i, because the possible proportionality function h, even if it is a unit, may grow fast at infinity.

The height T_f is independent of the homogeneous coordinates, because if we change $g_0, \ldots, g_n$ by a homogeneous factor, i.e. multiply by a meromorphic function h as above, then all the max in all the terms of the formula change by $\log\|h\|_v$, where v ranges over v_θ, v_a, and $-\log|c_h|$. The sum of these is 0 by Jensen's formula, i.e. by the sum formula. This definition goes back to Cartan [Ca 2], who also observed:

Suppose we have a single meromorphic function f, which defines a holomorphic map with projective coordinates $(1, f)$ in $\mathbf{P}^1$. Then

$$T_{(1,f)} = T_f - \log^+|c_f|,$$

where T_f is the previously defined Nevanlinna height of f.

This is immediate from the definitions, for instance writing $f = f_1/f_0$ where f_1, f_0 are entire without common zero, and applying the definition in this case, where the sum over $a \in \mathbf{D}_r$ is absent.

If f and $g: \mathbf{C} \to \mathbf{P}^n$ are two holomorphic maps and $g = A \circ f$ where A is a projective linear transformation, then

$$T_f = T_g + O(1).$$

This is immediate from the definitions. The $O(1)$ depends only on the linear transformation, not on f or g.

The next lemma shows that the heights of coordinate functions are bounded by the height of the mapping.

Lemma 3.1. *Let $f = (f_0, \ldots, f_n)$ with $f_0, \ldots, f_n$ entire without common zeros. Then*

$$N_{f_i}(r, 0) \leqq T_f(r) + O(1) \qquad \text{for} \quad r \to \infty.$$

If $g = (1, g_1, \ldots, g_n)$ with $g_i = f_i/f_0$ then

$$T_{g_i} \leqq T_f + O(1) \quad \text{for each } i.$$

Proof. By definition, looking at $(1, f_1/f_0, \ldots, f_n/f_0)$ we get:

$$T_f(r) = \int_0^{2\pi} \log \max_i \| f_i/f_0 \|_{r,\theta} \, \frac{d\theta}{2\pi} + \sum_{a \in \mathbf{D}_r} \log \max \| f_i/f_0 \|_{a,r} + O(1).$$

Since $f_0, \ldots, f_n$ are assumed relatively prime, $f_1/f_0, \ldots, f_n/f_0$ have poles where f_0 has zeros, and the sum over a is precisely $N_{f_0}(r, 0)$. Since the first integral is $\geqq 0$, we get $N_{f_0} \leqq T_f + O(1)$, which proves the first statement.

The second statement about T_{g_i} is immediate, because the max in the definition of the right-hand side involves more terms than the max in the definition of the left-hand side.

Observe that we cannot replace N_{f_i} by T_{f_i} in the first inequality, because $f_0, \ldots, f_n$ may be changed by a common factor h without zeros and poles, and h may grow arbitrarily fast. Such h disappears when we dehomogenize the coordinates, and consider f_i/f_0, for instance.

The height T_f of a map is the natural one. In §1, as a makeshift, we used the sum

$$T = T_{g_1} + \cdots + T_{g_n}.$$

From Lemma 3.1 we see that they have the same order of magnitude, since the inequality $T_f \leq T + O(1)$ is immediate.

Finally we give a relation which will be applied at once to the Fermat case.

Let $f = (f_0, \ldots, f_n)$ and $g = (f_0^d, \ldots, f_n^d) = f^d$, where d is a positive integer. Then

$$T_g = dT_f.$$

This is immediate from the definitions. Furthermore, let

$$f: \mathbf{C} \to \mathbf{P}^n \qquad \text{and} \qquad g: \mathbf{C} \to \mathbf{P}^m$$

be two maps into projective space, with coordinates $(f_0, \ldots, f_n)$ and $(g_0, \ldots, g_m)$. For $N = (n+1)(m+1) - 1$ we define the map

$$f \otimes g: \mathbf{C} \to \mathbf{P}^N \qquad \text{as having coordinates} \quad (f_i g_j).$$

Then

$$T_{f \otimes g} = T_f + T_g.$$

This is also immediate. However, in the application of this chapter, all we need is the formula with f^d.

VII, §4. THE FERMAT HYPERSURFACE

Having laid down these general notions of the height for maps into projective space, we come to the particular case when the image is contained in certain subsets.

The **Fermat variety** X in $\mathbf{P}^n$, of degree d, is defined by the equation

$$x_0^d + \cdots + x_n^d = 0.$$

One way to get a non-constant holomorphic map $f: \mathbf{C} \to X$ into the Fermat variety is to pick non-zero constants $c_0, \ldots, c_k$ with $k \geq 1$ such that

$$c_0^d + \cdots + c_k^d = 0,$$

and then for any non-constant holomorphic function h, and a constant $c \neq 0$ such that $c^d = -1$, we get a non-constant $\mathbf{C} \to \mathbf{P}^n$ by

$$(c_0 h, \ldots, c_k h, 1, c, 0, \ldots, 0).$$

The next theorem says that when the degree d is sufficiently large with respect to n, then this is essentially the only way. The reason for this will again come from the projective map given by the Wronskians, and we shall look explicitly into their "degrees" to obtain the desired inequality.

Theorem 4.1 (Green [Gr 2]). *Let $f = (f_0, \ldots, f_n): \mathbf{C} \to \mathbf{P}^n$ be a holomorphic map into the Fermat variety X, so*

$$f_0^d + \cdots + f_n^d = 0.$$

Suppose none of the f_i are 0. Define an equivalence $i \sim j$ if f_i/f_j is constant. If $d \geq n^2$ then for each equivalence class S we have

$$\sum_{i \in S} f_i^d = 0.$$

Proof. Instead of the Fermat equation as given, it is convenient for induction to use an equation

$$a_0 f_0^d + \cdots + a_n f_n^d = 0$$

with non-zero constants $a_0, \ldots, a_n$. Then by induction it suffices to prove that if $n \geq 2$, there exists a proper subset I of $\{0, \ldots, n\}$ such that the functions $\{f_i^d\}$ with $i \in I$ are linearly dependent. On the other hand, given the equation with the extra constants, after multiplying the functions f_i with a suitable constant (namely $a_i^{1/d}$) we get the ordinary Fermat equation without constants.

We let $g_i = f_i/f_0$ and $g = (1, g_1, \ldots, g_n)$. Then

$$g_1^d + \cdots + g_n^d = -1.$$

If $n \geq 2$, we have to show that $g_1^d, \ldots, g_n^d$ are linearly dependent. Suppose this is not the case. Then for each i,

$$g_i^d = L_i(g_1^d, \ldots, g_n^d)/L_0(g_1^d, \ldots, g_n^d) = L_i/L_0,$$

where for instance

$$L_0 = \begin{vmatrix} 1 & \cdots & 1 \\ (g_1^d)'/g_1^d & \cdots & (g_n^d)'/g_n^d \\ \vdots & & \vdots \\ (g_1^d)^{(n-1)}/g_1^d & \cdots & (g_n^d)^{(n-1)}/g_n^d \end{vmatrix},$$

and similarly for the other determinants L_i, $i = 1, \ldots, n$. Then we have two representations of the same projective map:

$$g = (1, g_1, \ldots, g_n), \qquad f = (f_0, \ldots, f_n),$$

and

$$L = (L_0, \ldots, L_n) \quad \text{is the same as } g^d \text{ and } f^d.$$

We then find by definition

$$dT_g(r) = T_L(r) = \int_0^{2\pi} \log \max_i \|L_i\|_{\theta,r} \frac{d\theta}{2\pi} + \sum_{a \in \mathbf{D}_r} \log \max_i \|L_i\|_{a,r} + O(1).$$

We get:

$$\int_0^{2\pi} \log \max_i \|L_i\|_{\theta,r} \frac{d\theta}{2\pi} \leqq \int_0^{2\pi} \log \max_i (1, \|L_i\|_{\theta,r}) \frac{d\theta}{2\pi}$$

$$\leqq \int_0^{2\pi} \sum_{i=0}^n \log^+ \|L_i\|_{\theta,r} \frac{d\theta}{2\pi}$$

$$\leqq \sum_{i=0}^n m_{L_i}(r)$$

[By Lemma 1.4] $\ll \log r + \sum_{i=1}^n \log^+ T_{g_i}(r)$

(1) $\ll \log r + \log^+ T_g(r)$

for $r \to \infty$ and r outside a set of finite measure. This last step uses Lemma 3.1. The implied constant depends on d and n as well as g. This takes care of the first term.

Next let us estimate the second term involving $a \in \mathbf{D}_r$. We have to estimate

$$N(r) = \sum_{a \in \mathbf{D}_r} \log \max_i \|L_i\|_{a,r}.$$

Consider for instance L_0. To simplify the notation, put

$$\varphi_i = g_i^d.$$

The only terms giving a positive contribution to this sum are the terms which are poles of L_0. The determinant has its usual expansion, and we get a pole of L_0 only at those a such that some term in the expansion has a pole. Consider for instance the term in this expansion given by

$$(\varphi_2^{(1)}/\varphi_2) \cdots (\varphi_n^{(n-1)}/\varphi_n).$$

Note that $\varphi_i^{(k)}/\varphi_i$ has a pole only at a zero of f_i, and that this pole has order at most $n - 1$. Hence

$$N(r) \leqq (n - 1) \sum_{i=0}^{n} N_{f_i}(r, 0)$$

$$\leqq (n - 1)(n + 1)T_g(r) + O(1) \qquad \text{(by Lemma 3.1).}$$

Hence we find

(2) $$\sum_{a \in \mathbf{D}_r} \max \log \|L_i\|_{a,r} \leqq (n^2 - 1)T_g(r) + O(1).$$

Putting together the computation at infinity and the computation for $a \in \mathbf{D}_r$. namely inequalities (1) and (2), we find

$$dT_g \leqq A \log r + B \log^+ T_g(r) + (n^2 - 1)T_g(r)$$

with constants A, B depending on d, n, f, and for r outside a set finite measure. If $d > n^2 - 1$, this implies that

$$T_g(r) = O(\log r)$$

for r outside a set of finite measure. But then all coordinates g_i $(i = 1, \ldots, n)$ are rational functions.

This implies that we can represent the rational map

$$f = (f_0, \ldots, f_n)$$

with polynomials $f_0, \ldots, f_n$ which have no common factor. Let e be the maximum degree of these polynomials.

Let us now use the Wronskians with f^d instead of g^d, so let

$$L_i = L_i(f^d) \qquad \text{for} \quad i = 0, \ldots, n,$$

where L_i is the logarithmic Wronskian as before, For instance, for $i = 0$, we have

$$L_0 = L_0(f^d) = \begin{vmatrix} 1 & \cdots & 1 \\ (f_1^d)'/f_1^d & \cdots & (f_n^d)'/f_n^d \\ \vdots & & \vdots \\ (f_1^d)^{(n-1)}/f_1^d & \cdots & (f_n^d)^{(n-1)}/f_n^d \end{vmatrix}$$

and similarly for all i where the omitted column is the i-th column. Then the projective maps

$$f^d = (f_0^d, \ldots, f_n^d) \quad \text{and} \quad L = (L_0, \ldots, L_n)$$

are equal. We shall compare degrees on both sides. For this we have to see a little better what $L_0, \ldots, L_n$ look like.

We use:

Let P be a polynomial of degree e. Then

$$(P^d)^{(k)} = P^{d-1}P_1 + \cdots + P^{d-k}P_k,$$

where $P_1, \ldots, P_k$ are polynomials with $\deg P_j \leq ej$.

This is immediate by induction. We then conclude that

$$\frac{(P^d)^{(k)}}{P^d} = \frac{P_1}{P} + \cdots + \frac{P_k}{P^k} = \frac{Q}{P^k} \quad \text{where} \quad \deg Q \leq ke.$$

We apply this remark to the determinant. Typically, the first term in the expansion of $L_0(f^d)$ can then be written in the form

$$\frac{Q_2}{f_2} \frac{Q_3}{f_3^2} \cdots \frac{Q_n}{f_n^{n-1}} = \frac{R}{f_0^{n-1} f_1^{n-1} \cdots f_n^{n-1}},$$

where R is a polynomial of degree $\leq (n-1)(n+1)e = (n^2-1)e$. In particular, the product $f_0^{n-1} f_1^{n-1} \cdots f_n^{n-1}$ is a common denominator for all the terms in all the expansions of all the determinants $L_0(f^d), \ldots, L_n(f^d)$. Hence we have an equality of projective maps

$$(f_0^d, \ldots, f_n^d) = (R_0, \ldots, R_n),$$

where $R_0, \ldots, R_n$ are polynomials of degree $\leq (n^2-1)e$. Now we can compare degrees. Since $f_0^d, \ldots, f_n^d$ are relatively prime, it follows that

$$de \leq (n^2-1)e.$$

This proves that $d \leq (n^2 - 1)$ unless $e = 0$, in which case f is a constant map. This contradiction concludes the proof of Green's theorem.

Theorem 4.1 can be applied to show that the Brody–Green perturbation of the Fermat hypersurface is hyperbolic.

For any complex number t we let X_t be the variety defined by

$$x_0^d + x_1^d + x_2^d + x_3^d + (tx_0x_1)^{d/2} + (tx_0x_2)^{d/2} = 0.$$

Thus X_0 is the Fermat variety of degree d. We take d even so. that the $d/2$ makes sense.

Theorem 4.2 (Brody–Green [B–G]). *If $d \geq 50$ is even, then X_t is hyperbolic for all but a finite number of $t \neq 0$.*

Proof. Since X_t is compact, by Brody's theorem it suffices to prove that X_t is Brody hyperbolic. Let

$$f: \mathbf{C} \to X_t, \qquad f = (f_0, \ldots, f_3)$$

be holomorphic. We have to show that f is constant. Let

$$
\begin{array}{ll}
g_0 = f_0^2, & g_3 = f_3^2, \\
g_1 = f_1^2, & g_4 = tf_0f_1, \\
g_2 = f_2^2, & g_5 = tf_0f_2.
\end{array}
$$

Then

$$g = (g_0, \ldots, g_5): \mathbf{C} \to \mathbf{P}^5$$

is holomorphic, and its image is contained in the Fermat hypersurface of degree $d/2$ in $\mathbf{P}^5$. It suffices to show that g is constant.

Case 1. Some $f_i = 0$. Then we may view f as a map into the curve obtained by setting $x_i = 0$ in the Brody–Green equation. For all but a finite number of t this is a non-singular curve of genus ≥ 2, so any holomorphic map of $\mathbf{C}$ into such a curve is constant.

Case 2. $f_i \neq 0$ for all i. By Green's Theorem 4.1, if $d/2 \geq 5^2 = 25$, so $d \geq 50$, then there is a partition of the set of indices $i = 0, \ldots, 5$ such that for each equivalence class S we have

$$\sum_{i \in S} g_i^{d/2} = 0.$$

And for $i, j \in S$ the function g_i/g_j is constant. It follows that each equivalence class has at least two elements.

Case 2(a). f_1/f_0 *and* f_2/f_0 *are both constant.* Then the original equation reduces to the form

$$af_0^d + f_3^d = 0$$

with some constant a, whence f_3/f_0 is constant, whence f is constant.

Case 2(b). f_2/f_0 *is constant but* f_1/f_0 *is not constant.* Then none of the ratios between g_0, g_1, g_4 can be constant. On the other hand, g_2/g_5 is constant. So 0, 2, 5 belong to the same class. But each class must have at least two elements, and the classes of 0, 1, 4 are distinct. Since we have only six elements to distribute among these three classes, the present case is impossible.

By symmetry, the case f_2/f_0 non-constant and f_1/f_0 constant is also excluded.

Case 2(c). Both f_2/f_0 and f_1/f_0 are not constant. As in the previous case, 0, 1, 4 lie in distinct clases. But 5 cannot be in the class of 0 (otherwise 2 is in this class), so the remaining 0-equivalence is that $3 \sim 0$, i.e. g_3/g_0 is constant. Thus we have the two possibilities:

(∗) $0 \sim 3, \qquad 1 \sim 2, \qquad 4 \sim 5,$

or

(∗∗) $0 \sim 3, \qquad 1 \sim 5, \qquad 2 \sim 4.$

From the equivalences (∗∗) we have by definition $f_0 f_1/f_2^2$ and $f_0 f_2/f_1^2$ constant. But then f_2^3/f_1^3 is constant, so f_2/f_1 is constant. This reduces to the equivalences (∗), in which case there are constants a, b such that

$$f_3 = af_0, \qquad f_2 = bf_1.$$

Then the original equation becomes

$$(1 + a^d)f_0^d + (1 + b^d)f_1^d + t^{d/2}(1 + b^{d/2})(f_0 f_1)^{d/2} = 0.$$

The last two coefficients cannot both vanish. Thus we obtain a non-trivial equation for f_1/f_0, which must therefore be constant, a contradiction which concludes the proof.

We also give Green's result concerning maps of $\mathbf{C}$ in the complement of Fermat varieties.

Theorem 4.3. (Green [Gr 2]). *Let* $f: \mathbf{C} \to \mathbf{P}^n$ *be a holomorphic map which omits the Fermat variety of degree* d. *Suppose* $d > n(n+1)$. *Let* $f = (g_0, \ldots, g_n)$ *with* $g_0 = 1$. *Define* $i \sim j$ *if* g_i/g_j *is constant. Then for each equivalence class* S *not corresponding to the constant functions, we have*

$$\sum_{i \in S} g_i^d = 0,$$

and for the class K *corresponding to the constant coordinates we have*

$$\sum_{k \in K} g_k^d \neq 0.$$

Proof. The proof is essentially the same as the previous theorem. Under the present assumption, write $f = (f_0, \ldots, f_n)$ where the f_i are entire without common zeros. Since f omits the Fermat hypersurface, we have

$$f_0^d + \cdots + f_n^d = h^d,$$

where h is entire without zeros, that is a unit. (We use here the fact that a unit u has a d-th root, namely $e^{(\log u)/d}$.) Consider the map

$$F: \mathbf{C} \to \mathbf{P}^{n+1} \quad \text{given by} \quad (f_0, \ldots, f_n, he^{i\pi/d}).$$

Then the image of F lies in the Fermat variety of degree d in $\mathbf{P}^{n+1}$. The following lemma concludes the proof.

Lemma 4.4. *If* $f_0, \ldots, f_n$ *are entire functions satisfying*

$$f_0^d + \cdots + f_n^d = h,$$

where h *is a unit, if* $f_0^d, \ldots, f_n^d$ *are linearly independent, and*

$$d > n(n+1),$$

then all the f_i/f_j *are constant.*

The proof is entirely similar to the proof of Theorem 4.1, and will be omitted.

As Green points out, if $d > n + 1$, then the image of f in either case (inside Fermat or omitting Fermat) should lie in some proper hypersurface section, which may however not be linear. Green gives an example of a holomorphic map

$$f: \mathbf{C} \to \mathbf{P}^2$$

omitting Fermat of degree 4, and whose image lies in a quartic but not in a line. The inequality $d > n + 1$ has to do with the canonical class of the Fermat variety. Cf. [La 3], which contains a general discussion of such matters.

VII, §5. ARBITRARY VARIETIES

By definition, varieties are assumed irreducible.

We want to formulate here the problem of Nevanlinna theory in the context of a holomorphic map

$$f: \mathbf{C} \to X$$

into a arbitrary projective variety X. The first thing to do is to define the height, counting functions, and proximity functions relative to X, and a divisor on X. The analogue of the First Main Theorem is true, and essentially immediate. The analogue of the Second Main Theorem is not known in general, and represents one of the basic unsolved problems of the field.

Let D be a Cartier divisor on X, represented locally for the Zariski topology by rational functions. Say on a Zariski open set U, D is represented by the function φ. This means $D|U = (\varphi)|U$.

By a **Weil function** for D we mean a function

$$\lambda_D: X - \operatorname{supp}(D) \to \mathbf{R}$$

which is continuous, and is such that if D is represented by φ on U, then there exists a continuous function $\alpha: U \to \mathbf{R}$ such that for all $x \notin \operatorname{supp}(D)$ we have

$$\lambda_D(x) = -\log|\varphi(x)| + \alpha(x).$$

The difference of two Weil functions is the restriction to $X - \operatorname{supp}(D)$ of a continuous function on X, and so is bounded. Thus two Weil functions differ by $O(1)$. It is easily shown that Weil functions exist. We show this later when we express them in terms of metrics on line bundles.

Suppose that $f(\mathbf{C})$ is not contained in D. This is equivalent with the fact that f meets D discretely, i.e. in any disc $\mathbf{D}_r$ there are only a finite number of points $a \in \mathbf{D}_r$ such that $f(a) \in D$. Given a Weil function λ_D associated with the divisor D, we define the **proximity function**

$$m_{f,D}(r) = m_f(r, D) = \int_0^{2\pi} \lambda_D(f(re^{i\theta})) \frac{d\theta}{2\pi}.$$

If D is effective, let $\lambda_D^+ = \max(0, \lambda_D)$. Then we could also use λ_D^+ instead of λ_D in this definition. Indeed, since λ_D becomes infinite in a neighborhood of D, and is otherwise bounded from below, λ_D is bounded from below, i.e. $\lambda_D \geq -O(1)$. If we put in the $+$ then we get another function differing from the above by $O(1)$. Adding a suitable positive constant to λ_D then makes this function positive. We are interested in these functions only mod $O(1)$. Since

$$\lambda_{D_1 + D_2} = \lambda_{D_1} + \lambda_{D_2} + O(1),$$

it follows that $m_f(r, D)$ is additive in D mod $O(1)$, that is

$$m_f(r, D_1 + D_2) = m_f(r, D_1) + m_f(r, D_2) + O(1),$$

assuming of course that f meets both D_1, D_2 discretely.

For any divisor D we let

$$D_f = f^{-1}(D),$$

as a divisor on $\mathbf{C}$, which exists by our basic assumption that f meets D discretely. Indeed, if D is represented by φ on U, and $f(a) \in U$, then we define

$$v_f(a, D) = v(a, D_f) = \mathrm{ord}_a(\varphi \circ f) \qquad (\geq 0 \text{ if } D \text{ is effective}),$$

$$N_{f,D}(r) = N_f(r, D) = \sum_{\substack{a \in \mathbf{D}_r \\ a \neq 0}} v_f(a, D) \log \left| \frac{r}{a} \right| + v_f(0, D) \log r.$$

We call $N_f(r, D)$ the **counting function**. It is the r-degree of $f^{-1}(D)$. (Compare with Chapter VI, §4.) It is additive in D, and for effective D it is ≥ 0 for $r \geq 1$ (the $\log r$ for $r < 1$ always interferes).

Finally we define the **height** (or **Cartan-Nevanlinna height**) to be

$$T_{f,D}(r) = T_f(r, D) = m_f(r, D) + N_f(r, D).$$

It follows immediately from the definitions that if $X = \mathbf{P}^n$ and D is a hyperplane section, say $x_0 = 0$ if $(x_0, \dots, x_n)$ are the projective coordinates, then $T_{f,D} = T_f + O(1)$, where T_f is the Cartan–Nevanlinna height. Also if $n = 1$, then $m_{f,D}$ and $N_{f,D}$ coincide up to $O(1)$ with the Nevanlinna functions denoted by the same letters. Then we get the **First Main Theorem**:

> **Theorem 5.1.** *If $D = (\varphi)$ is linearly equivalent to 0, that is D is the divisor of a function, then $T_f(r, D) = O(1)$, i.e. $T_f(r, D)$ is a bounded function of r for $r \to \infty$.*

Proof. We view $\varphi: X \to \mathbf{P}^1$ as a rational map of X into $\mathbf{P}^1$. Suppose first that f meets (φ) discretely. Then $\varphi \circ f$ is a meromorphic map $\mathbf{C} \to \mathbf{P}^1$. Furthermore, directly from the definitions we have

$$T_f(r, D) = T_{\varphi \circ f}(r, (0) - (\infty)) + O(1),$$

where $T_{\varphi \circ f}$ is the Cartan–Nevanlinna height of $\varphi \circ f$, which is equal up to $O(1)$ to the Nevanlinna height of a meromorphic function. Applying Proposition 2.1 of Chapter VI (Nevanlinna's First Main Theorem for meromorphic functions) concludes the proof in this case.

In general, let D be any Cartier divisor. There exists a rational function ψ on X such that $f(\mathbf{C})$ is not contained in $D + (\psi)$. The above argument applies, and allows us to define $T_f(r, D)$ for any Cartier divisor D, even if f meets D improperly. This concludes the proof that $T_f(r, D)$ depends only on the linear equivalence class of D.

The **Second Main Theorem** is today only a conjecture.

> **Conjecture 5.2.** *Let D be a divisor with simple normal crossings on the non-singular projective variety X. There exists a proper algebraic subset Z_D having the following property. Let $f: \mathbf{C} \to X$ be a holomorphic map such that $f(\mathbf{C}) \not\subset Z_D$. Let K be the canonical class, and let E be an ample divisor. Then*
>
> $$m_f(r, D) + T_f(r, K) \leqq O_{\text{exc}}(\log r + \log T_f(r, E)).$$

Only very special cases of this conjecture are known today, mostly for maps into $\mathbf{P}^n$ or Grassmannians, in a linear situation. In the next section, we give the typical example of what is known, in the form of a theorem due to Cartan, with an improvement in the formulation. Indeed, a term corresponding to the ramification term is missing in the above conjecture, but should be present in line with the $N_{1,f}$ term in Nevanlinna's Second Main Theorem for meromorphic functions in Chapter VI. We shall give such a term in a special, linear case, at the cost of imposing a linear non-degeneracy condition on the map f, so we don't get the set Z_D as above.

Aside from the first work of Cartan, extended to the derived curves by Weyl and Ahlfors, there is Fujimoto's treatment of the derived curves by Cartan's method [Fu 5], as well as the work of Stoll and Wu. See for instance the references in the bibliography. For the differential geometric treatment with hyperbolic forms, see Cowen–Griffiths, and Stoll [Stoll 2]. Noguchi [No 4] treats an important non linear special case involving abelian varieties. Cowen–Griffiths, followed by Noguchi, deal with maps $f: Y \to X$ such that dim $Y \geq X$ and f has maximal rank at some point. The case when dim $Y <$ dim X, especially when $Y = \mathbf{C}$, is much more subtle. For some attempts in this direction, see for instance [No 5], and the last corollary in [Ax]. For the general equidimensional case, see Carlson–Griffiths. For the arithmetic translation and point of view, see Vojta.

The strength of the conjecture can be seen from the special case when $D = 0$ and K is ample. In that case, the inequality

$$T_{f,K}(r) = O_{\mathrm{exc}}(\log r)$$

implies that f is algebraic, so degenerate when dim $X > 1$. Not even this special case is known (conjecture of Green–Griffiths), let alone Conjecture 5.6 of Chapter IV.

The determination of the set Z_D both in general and in concrete cases is a basic and interesting problem, with its own structure.

We conclude this section by general remarks concerning the height functions. We may summarize some of their properties as follows.

Let $f: \mathbf{C} \to X$ be a holomorphic map into a projective variety. To each Cartier divisor D on X one can associate a function $T_{f,D}$ of real numbers ≥ 1, depending only on the linear equivalence class of D, and uniquely determined up to a bounded function by the following properties:

H 1. *The map $D \mapsto T_{f,D}$ is a homomorphism* mod $O(1)$.

H 2. *If E is very ample, and $\psi: X \to \mathbf{P}^n$ is one of the associated imbeddings into projective space, then*

$$T_{f,E} = T_{\psi \circ f} + O(1),$$

where $T_{\psi \circ f}$ is the Cartan–Nevanlinna height.

We have proved the existence, and the uniqueness follows from the fact that given any Cartier divisor, D, there exists a very ample divisor E such that $D + E$ is very ample (an elementary fact of algebraic geometry). That $T_{f,D}$ depends only on the linear equivalence class of D is the geometric version of Propositions 2.1 and 2.2 in Chapter VI.

Vojta first pointed out the analogy between Nevanlinna theory and the theory of heights in algebraic number theory [Vo]. We have listed two of the standard properties of heights above. The others are equally true, with the same proofs. For instance:

H 3. *For any Cartier divisor D and ample E we have*

$$T_{f,D} = O(T_{f,E}).$$

Indeed, given $-D$ there exists a multiple nE such that nE is very ample and $-D + nE = E'$ is very ample. Then

$$-T_{f,D} + T_{f,nE} = T_{f,E'} \geq -O(1).$$

This yields $T_{f,D} \leq nT_{f,E} + O(1)$. Applying the argument to $-D$ instead of D yields the desired property.

H 4. *If D is effective, and $f(\mathbf{C}) \not\subset D$, then $T_{f,D} \geq -O(1)$.*

This comes for instance from the construction of $T_{f,D}$ given at the beginning of the section, since both terms $m_f(r, D)$ and $N_f(r, D)$ are positive (up to a bounded term).

H 5. *The association $(f, D) \mapsto T_{f,D}$ is functorial in (X, D). In other words, if $\psi : X \to Y$ is a morphism of varieties, and $D = \psi^{-1}D'$ where D' is a divisor on Y, then*

$$T_{f,D} = T_{\psi \circ f, D'} + O(1).$$

The proof is immediate from the definitions. If we combine **H 5** with **H 2** then we obtain a strengthened version, which is very useful in practice, namely:

H 6. *Let $\psi : X \to \mathbf{P}^M$ be a morphism, and suppose $D = \psi^{-1}(H)$ where H is a hyperplane. Then*

$$T_{f,D} = T_{\psi \circ f} + O(1),$$

where $T_{\psi \circ f}$ is the Cartan–Nevanlinna height.

Other properties of the height can be copied for instance from [La 5], Chapter 4, §1, §2, §3.

Appendix

Although we shall not use this appendix in the sequel, I thought valuable to have it here for comparison with other literature. I chose one definition of the height. A more differential geometric definition can be given as follows. Let L be a line bundle over X. Let s be a meromorphic section whose divisor is D. Let there be a given metric ρ on L, and write $|\ | = |\ |_\rho$. Then

$$\lambda_D(x) = -\log|s(x)|$$

is a Weil function associated with D. Similarly, we define the **proximity function** by

$$m_f^\rho(r, s) = \frac{1}{2}\int_{S_r} -\log|s \circ f|^2 \sigma,$$

where $\sigma = d\theta/2\pi$ with respect to the variable θ. Then

$$m_f^\rho(r, s) = m_f^\rho(r, s_1) + O(1) = m_f(r, D) + O(1),$$

if s, s_1 are two meromorphic sections having the same divisor D, and $m_f(r, D)$ is the proximity function defined at the beginning of the section in the Nevanlinna style.

Next for the height, we shall use the standard symbols

$$d = \partial + \bar\partial \qquad \text{and} \qquad d^c = \frac{1}{2\pi}\frac{\partial - \bar\partial}{2i}$$

In one variable $z = re^{i\theta}$ we have

$$d^c = \frac{1}{4\pi}r\frac{\partial}{\partial r} \otimes d\theta - \frac{1}{4\pi}\frac{1}{r}\frac{\partial}{\partial\theta} \otimes dr.$$

We let $\mathbf{S}_r$ be the circle of radius r.

Given a holomorphic map

$$f: \mathbf{C} \to X,$$

let ω be a real $(1, 1)$-form on X. We define

$$A_f^\omega(r) = \int_{\mathbf{D}_r} f^*\omega,$$

and the corresponding **Ahlfors–Shimizu height**

$$T_f^\omega(r) = \int_0^r A_f^\omega(t)\frac{dt}{t}.$$

The $(1, 1)$-form is usually that associated with a metric on a line bundle. Let L be a line bundle with a hermitian metric ρ. Let

$$\omega = c_1(\rho).$$

If s is locally a holomorphic section of L, then locally ω is given by

$$\omega = -dd^c \log|s|^2.$$

Two metrics differ multiplicatively by a positive C^∞ function u, bounded away from 0 and ∞. Hence a different choice of metric gives rise to another $(1, 1)$-form $\omega' = c_1(\rho')$ such that

$$\omega' - \omega = dd^c\gamma,$$

where $\gamma = \log u$ is a C^∞ function, which is bounded. Hence the two corresponding heights differ additively by

$$\int_0^r \int_{\mathbf{D}_t} dd^c(\gamma \circ f)\frac{dt}{t} = \int_0^r \int_{\mathbf{S}_t} d^c(\gamma \circ f)\frac{dt}{t} \qquad \text{(Stokes theorem)}$$

$$= \int_0^r \frac{1}{2}t\frac{\partial}{\partial t}\int_0^{2\pi} (\gamma \circ f)(te^{i\theta})\frac{d\theta}{2\pi}\frac{dt}{t}$$

$$= \frac{1}{2}\int_0^{2\pi} (\gamma \circ f)(re^{i\theta})\frac{d\theta}{2\pi} - \tfrac{1}{2}(\gamma \circ f)(0)$$

$$= O(1)$$

because γ is a bounded function. Thus the Ahlfors–Shimizu height is well defined by the line bundle mod $O(1)$.

Suppose our map $f: \mathbf{C} \to \mathbf{P}^n$ is into projective space, and let

$$\omega = dd^c \log \|z\|^2 \qquad \text{where} \qquad \|z\|^2 = \sum_{i=0}^n z_i\bar{z}_i.$$

Then ω is called the **Fubini–Study form**. We shall now prove that the Cartan–Nevanlinna height and the Ahlfors–Shimizu height differ by $O(1)$. For estimating purposes, they can be used interchangeably, according to which is more useful. In other words:

Proposition 5.3. *Let ω be the Fubini–Study form. Then*

$$T_f^\omega = T_f + O(1).$$

Proof. The proof is again by Stokes' theorem. Represent

$$f = (f_0, \ldots, f_n)$$

where the f_i are entire without common zero. Then:

$$
\begin{aligned}
T_f^\omega(r) &= \int_0^r \int_{\mathbf{D}_t} dd^c \log \| f(z) \|^2 \frac{dt}{t} \\
&= \int_0^r \int_{\mathbf{S}_t} d^c \log \| f(z) \|^2 \frac{dt}{t} \\
&= \int_0^r \frac{1}{4\pi} t \frac{\partial}{\partial t} \int_0^{2\pi} \log \| f(te^{i\theta}) \|^2 \, d\theta \frac{dt}{t} \\
&= \frac{1}{4\pi} \int_0^{2\pi} \log \| f(re^{i\theta}) \|^2 \, d\theta - \frac{1}{4\pi} \int_0^{2\pi} \log \| f(0) \|^2 \, d\theta,
\end{aligned}
$$

so we finally obtain

$$T_f^\omega(r) = \frac{1}{2} \int_0^{2\pi} \log \| f(re^{i\theta}) \|^2 \frac{d\theta}{2\pi} + O(1),$$

and by definition

$$\| f(re^{i\theta}) \|^2 = \sum_{j=0}^n |f_j(re^{i\theta})|^2.$$

Let $u(z) = \log \max_j |f_j(z)|$. Since the sup norm is equivalent to the euclidean norm (each is less than a positive constant times the other, and the positive constant is easily explicitly given), it follows that

$$2u(z) = \log \| f(z) \|^2 + O(1),$$

where in fact here, the $O(1)$ depends only on n (and could be taken as $\log n$). Using the definition of T_f at the beginning of §3, we see that

$$T_f^\omega(r) = T_f(r) + O(1),$$

thus proving the proposition.

If ω is a positive $(1, 1)$-form (as is the Fubini–Study form) then the definition of T_f^ω as an integral involving $f^*\omega$ makes it immediately clear that $T_f^\omega(r)$ is an increasing function of r. With the Nevanlinna definition of Chapter VI, one had to give an argument of a few lines to express T_f as an integral of a non-negative function.

VII, §6. SECOND MAIN THEOREM FOR HYPERPLANES

This entire section is due to Cartan [Ca 2], except as noted.

Theorem 6.1. *Let* $f = (f_0, \ldots, f_n): \mathbf{C} \to \mathbf{P}^n$ *be a holomorphic map, with* $f_0, \ldots, f_n$ *entire without common zeros. Assume that the image of f is not contained in any hyperplane. Let* $H_1, \ldots, H_q$ *be hyperplanes in general position. Let* $W(f) = W(f_0, \ldots, f_n) = W$ *be the Wronskian. Then*

$$\sum_{k=1}^{q} m_f(r, H_k) + T_f(r, K) + N_W(r, 0) \leq S(r),$$

where $S(r) = O_{\text{exc}}(\log r + \log^+ T_f(r))$ *is the usual Nevanlinna error term, and* $K = -(n+1)H$ *for any hyperplane H.*

The formula in the final conclusion is an improvement on Cartan's formulation. The difference is that Cartan does not give the term $N_{1/W}$ which depends only on the mapping f, but has a term which seems to depend on the chosen hyperplanes. This is the "wrong" structure for this term, which should reflect only the ramification of the mapping f, independently of the divisor D, which in this case is $\sum H_k$. At the end of the proof, we adjust Cartan's proof to obtain the result as stated.

As a matter of notation, we shall also denote by $H_1, \ldots, H_q$ the linear forms which define the hyperplanes. In particular, if $x = (x_0, \ldots, x_n)$ are homogeneous variables of $\mathbf{P}^n$, then we may form $H_k(x)$.

The rest of this section is devoted to the proof.

Note that we originally defined "general position" for more than $n+1$ hyperplanes. But if $q \leq n+1$ we can define **general position** to mean simply linearly independent. Any q linearly independent hyperplanes can be extended to more than $n+2$ hyperplanes in general position, and the theorem with $q \geq n+2$ implies the theorem with $q \leq n+1$. So without loss of generality, we assume $q \geq n+2$.

First we consider a certain morphism of $\mathbf{P}^n$. Let

$$n + 1 + p = q \qquad \text{so} \qquad p = q - (n+1).$$

Let $K \subset \{1, \ldots, q\}$ with $|K| = p$, say $K = \{k_1, \ldots, k_p\}$. We denote

$$H_K(x) = H_{k_1}(x) \cdots H_{k_p}(x) \qquad \text{for} \qquad x \in \mathbf{P}^n.$$

By a simple direct argument which we leave to the reader, one sees that

$$\psi: x \mapsto (\ldots, H_K(x), \ldots)_{|K|=p}$$

gives a morphism $\psi: \mathbf{P}^n \to \mathbf{P}^M$ (with suitable M), that is given x there exists K such that $H_K(x) \neq 0$. But we have the linear equivalence $H_K \sim pH_k$ for each of the indices K and k, and H_K is the inverse image under ψ of a hyperplane of $\mathbf{P}^M$, so by property **H 6** we get

$$pT_f = T_{\psi \circ f} + O(1).$$

(*Remark*: Cartan gives a simple ad hoc estimate at this point, while we use the general principle given by **H 6**.)

We let

$$v(z) = \log \max_{|K|=p} |H_K(f(z))|.$$

By definition,

$$T_{\psi \circ f}(r) = \int_0^{2\pi} v(re^{i\theta}) \frac{d\theta}{2\pi} + O(1).$$

We let $W(f_0, \ldots, f_n)$ be the Wronskian. For each $(n+1)$-tuple $I = \{i_1, \ldots, i_{n+1}\}$ among $\{1, \ldots, q\}$ if c_I is the determinant of the linear transformation going from the standard coordinates to $H_{i_1}, \ldots, H_{i_{n+1}}$, then

$$c_I W(h_{i_1}, \ldots, h_{i_{n+1}}) = W(f_0, \ldots, f_n).$$

As in the previous sections, starting with §1, we let

$$L_I = L(h_{i_1}, \ldots, h_{i_{n+1}}) = W(h_{i_1}, \ldots, h_{i_{n+1}})/h_{i_1} \cdots h_{i_{n+1}},$$

so L_I is the determinant formed with the higher logarithmic derivatives. We shall use the following

Basic Formula. *Let* $\{1, \ldots, q\} = \{i_1, \ldots, i_{n+1}, k_1, \ldots, k_p\}$ *be decomposed into two complementary families of indices. Let*

$$G = h_1 \cdots h_q / W(f_0, \ldots, f_n).$$

Then

$$h_{k_1} \cdots h_{k_p} = G c_I L(h_{i_1}, \ldots, h_{i_{n+1}}).$$

The formula is obvious from the definitions, but it gives an expression for the coordinate $H_K(f)$ on the left in terms of something which does not depend on K, namely the quotient $h_1 \cdots h_q / W(f_0, \ldots, f_n)$, and a term

$L(h_{i_1}, \ldots, h_{i_{n+1}})$ which turns out to have a very small height, as we well know from the preceding sections. Indeed, we get

$$v = \log|G| + O(1) + \max_I \log|L(h_{i_1}, \ldots, h_{i_{n+1}})|,$$

and we claim that

$$\int_0^{2\pi} v(re^{i\theta}) \frac{d\theta}{2\pi} \leqq \int_0^{2\pi} \log|G(re^{i\theta})| \frac{d\theta}{2\pi} + O_{exc}(\log r + \log T_f(r)).$$

To justify the O_{exc} term, let T_I be the height of the map $(h_{i_1}, \ldots, h_{i_{n+1}})$, and let $L_I = L(h_{i_1}, \ldots, h_{i_{n+1}})$. Then

$$\int_0^{2\pi} \max_I \log|L(h_{i_1}, \ldots, h_{i_{n+1}})(re^{i\theta})| \frac{d\theta}{2\pi}$$

$$\leqq \sum_I \int_0^{2\pi} \log^+ |L(h_{i_1}, \ldots, h_{i_{n+1}})(re^{i\theta})| \frac{d\theta}{2\pi}$$

$$= \sum_I m_{L_I}(r)$$

$$\leqq \sum_I O_{exc}(\log r + \log T_I(r)) \qquad \text{by Lemma 1.4}$$

$$\leqq O_{exc}(\log r + \log T_f(r))$$

because for each I, the maps f and $(h_{i_1}, \ldots, h_{i_{n+1}})$ differ by a projective linear transformation.

There remains to estimate $\int \log|G|$, and this is where we adjust Cartan's proof. We recall that for any meromorphic function g,

$$\int \log|g| = \int \log^+ |g| - \int \log^+ |1/g|$$

$$= m_g - m_{1/g}$$

$$= N_{1/g} - N_g + O(1).$$

We apply this to

$$G = h/W, \qquad \text{where} \quad h = h_1 \cdots h_q.$$

We then find

$$\int \log|G| = \sum_{k=1}^q N_{1/h_k} - \sum_{k=1}^q N_{h_k} - N_{1/W} + N_W + O(1).$$

But h_k, W are entire functions so $N_{h_k} = N_W = 0$. Also $N_{1/h_k} = N_f(H_k)$. Hence from Property **H6** we obtain

$$pT_f \leqq \sum_{k=1}^{q} N_f(H_k) - N_{1/W} + S,$$

where S is the error term. Since $p = q - (n + 1)$, the left-hand side may be written

$$pT_f = \sum_{k=1}^{q} m_f(H_k) + \sum_{k=1}^{q} N_f(H_k) - (n + 1)T_f(H) + O(1).$$

We now cancel the sum $\sum N_f(H_k)$ on each side to conclude the proof. Note that the formula

$$\boxed{(q - n - 1)T_f \leqq \sum_{k=1}^{q} N_f(H_k) - N_{1/W} + S}$$

could already be regarded as an alternative statement of the theorem.

We note that Borel's theorem is a special case of Cartan's theorem. Formally we have the following corollary.

Corollary 6.2. *Let* $g_1, \ldots, g_n$ *be entire functions without zeros (so units in the ring of entire functions). Suppose that*

$$g_1 + \cdots + g_n = 1.$$

Then $g_1, \ldots, g_n$ *are linearly dependent if* $n \geq 2$.

Proof. Let $g: \mathbf{C} \to \mathbf{P}^{n-1}$ be the map $(g_1, \ldots, g_n)$. Let $x_1, \ldots, x_n$ be the homogeneous variables of $\mathbf{P}^{n-1}$. Let H_k be the hyperplane $x_k = 0$ for $k = 1, \ldots, n$ and $x_1 + \cdots + x_n = 0$ for $k = n + 1$. Thus g does not meet these hyperplanes. Hence

$$m_g(r, H_k) = T_g(r, H_k) + O(1)$$

for $k = 1, \ldots, n + 1$. The canonical class is the class of $-nH$ for any hyperplane H. If $g_1, \ldots, g_n$ are linearly independent, then by Cartan's theorem

$$(n + 1)T_g \leqq nT_g + O_{\text{exc}}.$$

Hence $T_g(r) = O_{\text{exc}}(\log r)$, and therefore all $g_1, \ldots, g_n$ are polynomials, so are constant since g_i is a unit for each i. This concludes the proof of the corollary.

CHAPTER VIII

Normal Families of the Disc in $\mathbf{P}^n$ Minus Hyperplanes

After the work of Nevanlinna, Bloch wrote a fundamental paper where, as he says (my translation):

> There was no doubt *a priori* that to Borel's theorem there should correspond theorems in finite terms, just as the Landau theorem corresponds to Picard's theorem. This results from a principle which can be stated as follows: *Nihil est in infinito quod non prius fuerit in finito.* But what were these theorems? This was not at all clear at first sight.

Indeed, in Nevanlinna theory and Borel's theorem, the results are formulated in terms of maps of $\mathbf{C}$ into something, whether into $\mathbf{P}^1$ itself or projective space. Bloch wanted to give explicit bounds to the radii of discs which could be mapped into the something. This is what corresponds to the Schottky–Landau theorem.

Let Y be a projective variety. Let Z be a proper Zariski closed subset. We are interested in whether $Y - Z$ is hyperbolic, and hyperbolically imbedded in Y. In addition, let S be another proper Zariski closed subset. We are interested in whether $Y - Z$ is hyperbolically imbedded in Y modulo S. Only very special cases are known. In this chapter, we reproduce results of Bloch and part of Cartan's thesis dealing with the problem, for $\mathbf{P}^n$ minus $n + 2$ hyperplanes, when the special set S consists of the diagonal hyperplanes. We suggest that the reader look at Cartan's conjecture in §4, and the statements of the known results in §2, §4, and §5, before looking at the proofs. This will give the reader a basic idea where the theory is going, and will show how to connect with Chapters II and III.

The techniques of proof used are again those of Bloch, but applied by Cartan to families of mappings of $\mathbf{D}$ into the variety instead of mapping all of $\mathbf{C}$ into the variety. As Bloch says, this theory belongs to the "finite" version of the Schottky–Landau theorems, and in contemporary language, to the "finite" version of Brody hyperbolicity (and so Kobayashi hyperbolicity).

Bloch proves that certain families of holomorphic maps of the unit disc into projective space minus some hyperplanes are normal. But his families are subject to certain restrictions, like fixing the value at the origin. At this point, Cartan goes substantially beyond Bloch, using essentially the same techniques, by considering more general families. Cartan himself says:

> The "translation in finite terms" of Borel's theorem, in the case of an arbitrary number of functions, was first considered in 1926 by A. Bloch in his paper which appeared in the *Annales de l'École Normale Supérieure*. This geometer succeeded in overcoming to a large extent the genuine difficulties raised by this question, and he obtained a theorem which is an immediate generalization of Schottky's theorem, but which is more complicated because of the existence of *singular cases*; these have only been partly cleared up.

And whereas Montel had proved that a family of holomorphic functions omitting three points in $\mathbf{P}^1$ is normal, thus giving new insight into Picard's theorem, Cartan goes on to say:

> After this work of A. Bloch, in the case of Borel's identity, there was still lacking the analogue of P. Montel's criterion in the case of an identity with three terms. This is the kind of theorem I have tried to prove, without actually using explicitly Bloch's results, which did not seem to me completely proved. I have found it useful to take up the question from the beginning. Even if I have not substantially modified the general ideas, which were in some sense in the nature of things, I have however oriented my proof in a slightly different direction, since A. Bloch had limited himself to consider systems of p functions *taking given values at the origin*. By getting rid of this restriction, I obtained a sort of *criterion for complex normal families...*

In the present chapter, we give an account of this work, and follow Cartan.

Since I am principally interested in the direct application to the hyperbolic imbedding of the complement of hyperplanes in projective space modulo the diagonals, I do not give all the results in Cartan's thesis [Ca 1], but only those directly relevant to the application. I strongly recommend referring to that paper for the additional results it contains.

Furthermore, except for the case of three functions, the case of four functions and a fortiori n functions is quite complicated. I have been unable to simplify Cartan's arguments. They appeal to higher logarithmic derivatives, which also occur in Ahlfors' study of the "derived" curves [Ah], see the exposition in Vojta [Vo], making the link with Schmidt's techniques in the number theory case. This is one aspect of the theory which still needs a lot of clarification and simplification. I hope that by reproducing certain parts of Cartan's thesis, I can draw attention to the importance of dealing with these problems.

Bloch, by the way, was very pleased with himself (and rightly so) as shown by the comments at the end of his paper, which I reproduce in the original tongue:

> Nous demanderons, en terminant, que l'on veuille bien excuser les moyens relativement compliqués par lesquels nous avons obtenu des résultats en somme assez simples, comme le théorème III; les démonstrations actuelles seront sans doute ultérieurement simplifiées, et l'on en pourra trouver d'autres. Mais actuellement, on doit reconnaitre que bon nombre des auteurs qui traitent ou croient traiter de la théorie des fonctions analytiques ont beaucoup plus tendance à suivre le caprice de leur fantaisie qu'à examiner les problèmes qui se posent en réalité, et ne produisent par suite que des oeuvres dont le caractère violemment inesthétique n'a d'égal que l'absolue stérilité. De tels écrits comptent assurément parmi les plus remarquables spécimens de ce que l'on peut appeler la "contre-mathématique".*

VIII, §1. SOME CRITERIA FOR NORMAL FAMILIES

We start *ab ovo* with the logarithmic derivative of a unit. We return to the considerations of Lemma 3.2 of Chapter VI.

Lemma 1.1. *Given a positive integer n, there exists a polynomial P_n with positive coefficients depending only on n having the following property. Let h be a unit on a disc $\bar{\mathbf{D}}_R$, and let $0 \leqq r < R$. Then for $|z| = r$ we have*

$$|h^{(n)}/h(z)| \leqq P_n\left(R, \frac{1}{R - r}, m_h(R, 0 + \infty)\right).$$

Proof. By Poisson–Jensen,

$$\log h(z) = \int_0^{2\pi} \log |h(Re^{i\theta})| \frac{Re^{i\theta} + z}{Re^{i\theta} - z} \frac{d\theta}{2\pi} + iC$$

---

* *Translation:* In conclusion, we ask to be excused for the relatively complicated means by which we have obtained results which are, after all, rather simple, such as theorem III; the present proofs will no doubt eventually be simplified, and one may find others. But today, one must recognize that a good number of authors who treat, or think they treat, the theory of analytic functions, have a much greater tendency to follow the whims of their fantasies than to examine the problems which arise in reality, and as a result, they produce works whose violently inaesthetic character is equalled only by their absolute sterility. Such writings assuredly count among the most remarkable specimens of what could be called "counter-mathematics".

and therefore

$$h'/h(z) = \int_0^{2\pi} \log |h(Re^{i\theta})| \frac{2Re^{i\theta}}{(Re^{i\theta} - z)^2} \frac{d\theta}{2\pi}.$$

We differentiate under the integral sign and obtain

$$\left(\frac{d}{dz}\right)^{n-1} (h'/h(z)) = n!(-1)^n \int_0^{2\pi} \log|h(Re^{i\theta})| \frac{2Re^{i\theta}}{(Re^{i\theta} - z)^{n+1}} \frac{d\theta}{2\pi},$$

so

$$\left|\left(\frac{d}{dz}\right)^{n-1} h'/h(z)\right| \leq \frac{n!2R}{(R - r)^{n+1}} [m_h(R) + m_{1/h}(R)].$$

For a unit h we have $m_h = T_h$ and $m_{1/h} = T_{1/h} = T_h - \log|c_h|$. The lemma follows at once.

If we apply the standard inequalities for m of a sum and product, then we deduce from Lemma 1.1 that

$$m_{h^{(n)}/h}(r) \leq A_n + B_n \log \frac{1}{R - r} + C_n \log^+ m_h(R, 0 + \infty) + D_n \log^+ R,$$

where A_n, B_n, C_n, D_n depend only on n. In the applications, we take R bounded, and even $R < 1$. Then the term $D_n \log^+ R$ vanishes, and we formulate the result as we shall need it.

Lemma 1.2. *There exist positive constants A_n, B_n, C_n depending only on n, having the following property. If h is a unit on $\bar{\mathbf{D}}_R$ and*

$$0 \leq r < R < 1$$

then

$$m_{h^{(n)}/h}(r) \leq A_n + B_n \log \frac{1}{R - r} + C_n \log^+ m_h(R, 0 + \infty).$$

Next we look at the application to the Borel equation.

Let $f_1, \ldots, f_n$ be holomorphic functions on $\mathbf{D}$. As before we let the Wronskian be

$$W(f) = W(f_1, \ldots, f_n) = \begin{vmatrix} f_1 & \cdots & f_n \\ f_1' & \cdots & f_n' \\ \vdots & & \vdots \\ f_1^{(n-1)} & \cdots & f_n^{(n-1)} \end{vmatrix}$$

and

$$L(f) = L_f = \frac{W(f_1,\dots,f_n)}{f_1\cdots f_n}.$$

We also let $L_{i,f}$ be the dehomogenized Wronskian as in Chapter VII, §1, that is

$$L_{i,f} = i\text{-th minor with respect to the first row of } L_f.$$

Lemma 1.3. *Let $\alpha > 0$ and let n be a positive integer. There exist constants A, B, C with A depending on n, α and B, C depending only on n having the following property. Let $f_1,\dots,f_n$ be units on $\mathbf{D}$ such that*

$$f_1 + \cdots + f_n = 1, \qquad |L_f(0)| > \alpha, \qquad |f_i(0)| > \alpha \qquad for \quad i = 1,\dots,n.$$

Let $m_f = m_{f_1} + \cdots + m_{f_n}$. Then for $0 \leq r < R < 1$ we have

$$m_f(r) \leq A + B \log \frac{1}{R-r} + C \log^+ m_f(R).$$

Proof. We can solve

$$f_i = L_i/L.$$

Then by Proposition 2.1 of Chapter VI

(∗)
$$m_{f_i} \leq m_{L_i} + m_{1/L}$$
$$\leq m_{L_i} + m_L - \log|L(0)|$$

and

(∗∗)
$$m_{f_i}(R, 0 + \infty) = m_{f_i}(R) + m_{1/f_i}(R)$$
$$= 2m_{f_i}(R) - \log|f_i(0)|.$$

Hence by Lemma 1.2,

$$m_f(r) \leq A_n + B_n \log \frac{1}{R-r} + C_n \sum \log^+ m_{f_i}(R, 0 + \infty) - n \log|L_f(0)|$$

$$\leq A_n + B_n \log \frac{1}{R-r} + C_n \sum \log^+ [m_{f_i}(R) - \tfrac{1}{2}\log|f_i(0)|]$$

$$- n \log|L_f(0)|.$$

by using (∗) and (∗∗). The lemma follows at once.

As in Chapter VI we want to get rid of R on the above estimates. We use the following variation on Borel's lemma, due to Cartan.

Lemma 1.4. *Let $S(r) \geq 0$ be an increasing function on an interval $a \leq r < b$. Let $\lambda > 0$. Then*

$$S(r + e^{-S(r)/\lambda}) \leq S(r) + \lambda \log 2$$

except for r in a set of measure $\leq 2e^{-S(a)/\lambda}$.

The proof is similar to that of Borel's Lemma 3.7 of Chapter VI, and will be omitted.

Lemma 1.5. *Let A, B, C be positive numbers. Let $S(r) \geq 0$ be an increasing function on $a \leq r < b$ satisfying*

$$S(r) \leq A + B \log \frac{1}{R - r} + C \log^+ S(R) \qquad for \quad a \leq r < R < b.$$

Then there exists $M > 0$ depending only on A, B, C such that for $a \leq r < b$ we have

$$S(r) \leq M \qquad or \qquad S(r) \leq 2B \log \frac{2}{b - r}.$$

Proof. Suppose that

$$S(r) > \lambda \log \frac{2}{b - r} \qquad \text{so that} \qquad 2e^{-S(r)/\lambda} < b - r.$$

Then there exists R with $r < R < b$ such that

$$S(R + e^{-S(R)/\lambda}) \leq S(R) + \lambda \log 2$$

by applying Lemma 1.4 to the points of the interval between R and b. Letting $\lambda = 2B$, we get

$$S(R) \leq A_1 + \frac{B}{\lambda} S(R) + C \log^+\left(S(R + e^{-S(R)/\lambda})\right)$$

$$\leq A_1 + \tfrac{1}{2}S(R) + C \log^+ S(R) + C \log^+(\lambda \log 2) + C \log 2,$$

whence

$$S(R) \leq A_2 + 2C \log^+ S(R)$$

and therefore

$$S(r) \leq S(R) \leq M$$

as desired.

Lemma 1.5 will of course be applied to the function $m_f(r)$ of Lemma 1.3. This will be done in the context of normal families, which we now enter.

Let $\mathscr{F}$ be a family of holomorphic functions on a domain U. We say that $\mathscr{F}$ is **normal** if, given an infinite sequence of functions in the family, there exists a subsequence which converges uniformly on every compact subset either to a holomorphic function or to the constant infinity.

As usual, we let $\mathbf{D}$ be the unit disc.

Theorem 1.6 (Montel). *Let $\mathscr{F}$ be a family of holomorphic functions on $\mathbf{D}$ such that $m_f(r)$ is bounded for $0 \leq r < 1$, and $f \in \mathscr{F}$. Then $\mathscr{F}$ is normal.*

Proof. We give the proof only in the case when $\mathscr{F}$ consists of units, that is invertible holomorphic functions, which is the only case to be used in the sequel. In this case, $m_f = T_f$ is an increasing function. Let

$$m_f(1) = \lim_{r \to 1} m_f(r).$$

Using $\log \alpha = \log^+ \alpha - \log^+ 1/\alpha$, we get the usual inequality from the Poisson-Jensen formula. For $z \in \mathbf{D}$ and $f \in \mathscr{F}$, and $|z| = r$,

$$\log|f(z)| \leq \frac{1+r}{1-r} m_f(1) - \frac{1-r}{1+r} m_{1/f}(1)$$

$$\leq \frac{1+r}{1-r} T_f(1) - \frac{1-r}{1+r} T_f(1) + \frac{1-r}{1+r} \log|f(0)|$$

by Proposition 2.1 of Chapter VI. Considering $1/f$ instead of f, we get the reverse inequality

$$-\frac{4r}{1-r^2} m_f(1) + \frac{1+r}{1-r} \log|f(0)| \leq \log|f(z)|$$

$$\leq \frac{4r}{1-r^2} m_f(1) + \frac{1-r}{1+r} \log|f(0)|.$$

Then for f denoting elements in a given sequence, either:

$|f(0)|$ is bounded, so $|f|$ is bounded, and the right-hand inequality together with the standard Montel theorem shows the uniform convergence on a disc of given radius < 1;

or $|f(0)|$ is unbounded, whence $|f|$ tends to ∞ on a disc of given radius < 1. This proves the theorem.

Proposition 1.7. *Let* $\alpha > 0$, *and let* n *be a positive integer. The set of solutions*

$$f_1 + \cdots + f_n = 1$$

in units $f_1, \ldots, f_n$ *on* $\mathbf{D}$ *such that*

$$|f_i(0)| > \alpha \quad for \ i = 1, \ldots, n \qquad and \qquad |L_f(0)| > \alpha$$

is a normal family. In fact, if $f: \mathbf{D} \to \mathbf{P}^n$ *is the map*

$$f = (1, f_1, \ldots, f_n)$$

then given $r < 1$, *the values* $m_f(r)$ *are bounded for all* f.

Proof. Immediate by putting together Theorem 1.6 and Lemmas 1.3, 1.5.

Theorem 1.8. *Let* $\mathscr{F}$ *be an infinite family of solutions of*

$$f_1 + \cdots + f_n = 1$$

in units $f_1, \ldots, f_n$ *on* $\mathbf{D}$. *Assume that there exists* $\alpha > 0$ *and* $0 < r_1 < 1$ *such that*:

(a)   *Given* $f \in \mathscr{F}$, *there exists* $z_1 \in \mathbf{D}_{r_1}$ *for which* $|L_f(z_1)| \geq \alpha$.
(b)   *Given* $f \in \mathscr{F}$ *and* i, *there exists* $z_2 \in \mathbf{D}_{r_1}$ *such that*

$$|f_i(z_2)| \geq \alpha.$$

Then $\mathscr{F}$ *is a normal family. In fact, given* $r < 1$, *the values* $m_f(r)$ *are bounded uniformly for* $f \in \mathscr{F}$.

Proof. Given a non-zero holomorphic function g on $\mathbf{D}$ let $z_1 \in \mathbf{D}_{r_1}$ be such that $g(z_1) \neq 0$. Fix r_2 with $r_1 < r_2 < 1$. Then from the Poisson–Jensen inequality, we get for $r_2 \leq r < 1$

$$\log|g(z_1)| \leq \frac{r + r_1}{r - r_1} m_g(r) - \frac{r - r_1}{r + r_1} m_{1/g}(r),$$

whence

$$m_{1/g}(r) \leq \left(\frac{r+r_1}{r-r_1}\right)^2 m_g(r) + \frac{r_2+r_1}{r_2-r_1} \log^+ \left|\frac{1}{g(z_1)}\right|$$

$$\leq \left(\frac{r_2+r_1}{r_2-r_1}\right)^2 m_g(r) + \frac{r_2+r_1}{r_2-r_1} \log^+ \left|\frac{1}{g(z_1)}\right|$$

[because $r \mapsto (r+r_1)/(r-r_1)$ is decreasing for $r > r_1$]

$$\leq K\left[m_g(r) + \log^+ \left|\frac{1}{g(z_1)}\right|\right],$$

where the constant K depends only on r_1 and r_2. So in (∗) and (∗∗), instead of the estimate for $m_{1/L}$ and m_{1/f_i} with $L(0)$ and $f_i(0)$, we use this more refined version to conclude the proof.

In later applications, we need another sufficient condition to get a normal family, besides the one furnished by Lemma 1.3. We give it in the next theorem. For this it is useful to adopt some notation. Let $J = (j_1, \ldots, j_q)$ be a subset of the indices $\{1, \ldots, n\}$. We let

$$L(J) = L(f_{j_1}, \ldots, f_{j_q}) \qquad \text{and} \qquad L(f) = L(1, \ldots, n).$$

Theorem 1.9. *Let $\mathscr{F}$ be an infinite family of solutions of*

$$f_1 + \cdots + f_n = 1$$

in units $f_1, \ldots, f_n$ on **D**. *Assume that there exists $r_1 < 1$ and a constant K such that for $f \in \mathscr{F}$ and $r_1 < r < 1$ we have*

$$m_{1/L(f)}(r) \leq K + K \sum_J m_{L(J)}(r),$$

where the sum is taken over some subsets J of $\{1, \ldots, n\}$. Then $\mathscr{F}$ is a normal family. In fact, given $r < 1$, the values $m_f(r)$ are bounded uniformly for $f \in \mathscr{F}$.

Proof. We write again $f_i = L_i/L$ so

$$m_{f_i} \leq m_{L_i} + m_{1/L}.$$

Instead of flipping $m_{1/L}$ as in Lemma 1.3 we merely apply the hypothesis directly together with Lemma 1.2 to get the same conclusion as in Lemma 1.3. Then Lemmas 1.4, 1.5 and Theorem 1.6 apply in the same way to conclude the proof.

In both Theorem 1.8 and 1.9 the problem is to estimate $m_{1/L}$. These two theorems give us the tools which will be used in §2, §4, and §5 under two different types of hypotheses, to conclude that a certain family is normal.

VIII, §2. THE BOREL EQUATION ON D FOR THREE FUNCTIONS

For the rest of this chapter, unless otherwise specified, by **convergence** *we mean uniform convergence on compact sets.*

We are interested in the relative local compactness of certain families of maps which are unit solutions of the Borel equation. For three functions, the proof is relatively simple, and uses only the preceding considerations and test for normality. Therefore we give it first in this section. We shall need additional lemmas and estimates to carry out the proof with more functions.

It will be useful to make one definition, to select out one of the possible convergences. Let $\mathscr{S}$ be a sequence of units on **D**. We say that $\mathscr{S}$ is **C*-convergent** if $\mathscr{S}$ converges to a unit on **D**. Thus the limit function has no zero or pole on **D**. We also recall here that if a sequence of holomorphic functions converges on some connected open set U in **C**, then the zeros of the limit function are limits of the zeros of the functions in the sequence, or the limit function is identically 0. This is immediate from the formula expressing the number of zeros as an integral of the logarithmic derivative.

Theorem 2.1. *Let $\mathscr{F} = \{f\}$ be an infinite family of solutions of Borel's equation*

$$f_1 + f_2 + f_3 = 0$$

with units f_1, f_2, f_3 on **D**. *Then $\mathscr{F}$ is relatively locally compact in* Hol(**D**, **P**²). *In fact, there exists a subsequence $\mathscr{S}$ of $\mathscr{F}$ having one of the following properties:*

(a)   *All quotients f_i/f_j are* **C*-***convergent for $f \in \mathscr{S}$.*
(b)   *For some pair of indices i, k and $f \in \mathscr{S}$ the sequence $\{f_k/f_i\}$ converges to 0; or equivalently, for some pair of indices $i \neq j$, the sequence $\{f_i/f_j\}$ is* **C*-***convergent to* -1.

Proof. Because of the possibility of taking diagonal sequences, it suffices to prove the convergence properties (a) or (b) on a disc of given radius $r_0 < 1$. Indeed, there is a countable sequence of radii approaching 1, and there are only four ways of renumbering the indices.

We now consider cases, following a pattern which will be reproduced in the generalizations to four and arbitrarily many functions.

Case 0.1. There exists a subsequence for which some quotient f_k/f_i converges to 0 on $\mathbf{D}_{r_0}$; or equivalently there exists a subsequence for which some quotient f_i/f_j is $\mathbf{C}^*$-convergent to -1 on $\mathbf{D}_{r_0}$.

The equivalence is immediate, for if, say, f_3/f_1 converges to 0, then from

$$1 + \frac{f_2}{f_1} + \frac{f_3}{f_1} = 0$$

we see that f_2/f_1 converges to -1, and conversely.

Case 0.2. There exists a subsequence $\mathscr{S}$ such that for $f \in \mathscr{S}$, some quotient f_i/f_j is $\mathbf{C}^*$-convergent to a limit $\neq -1$ on $\mathbf{D}_{r_0}$.

Say the quotient is f_2/f_1. Then

$$\frac{f_3}{f_1} = -\left(\frac{f_2}{f_1} + 1\right)$$

is also $\mathbf{C}^*$-convergent, and finally f_2/f_3 is $\mathbf{C}^*$-convergent, and condition (a) is satisfied.

Case 1. Suppose Case 0.1 does not hold. Then there exists $\alpha > 0$ such that for all quotients f_i/f_j, $f \in \mathscr{F}$, there exist $z_1, z_2 \in \mathbf{D}_{r_0}$ satisfying

$$\alpha \leqq |f_i/f_j(z_1)| \quad \text{and} \quad |f_i/f_j(z_2)| \leqq 1/\alpha.$$

Case 1.1. There exists an infinite sequence $\mathscr{S}$ in $\mathscr{F}$ for which the log derivative of, say, f_2/f_1 is bounded on $\mathbf{D}_{r_0}$. Then in this case, f_2/f_1 is bounded, and hence the family of functions $\{f_2/f_1\}_{f \in \mathscr{S}}$ is normal.

This comes from a general result:

Lemma 2.2. *Let $\{h\}$ be a family of units on $\mathbf{D}_{r_0}$ such that $|h'/h|$ is bounded on $\mathbf{D}_{r_0}$. Assume that for each h there exist $z_1, z_2 \in \mathbf{D}_{r_0}$ such that $\alpha \leqq |h(z_1)|$ and $|h(z_2)| \leqq 1/\alpha$. Then $\{h\}$ is bounded, so $\{h\}$ is a normal family.*

Proof. Write $h = e^g$ where $|g'|$ is bounded by B, say, on $\mathbf{D}_{r_0}$. By the Mean Value Theorem, for $z, z' \in \mathbf{D}_{r_0}$ we have

$$|g(z) - g(z')| \leqq B|z - z'| \leqq 2B.$$

Since $|e^w| \leq e^{|w|}$ for complex w, we find

$$|e^{g(z)}/e^{g(z')}| = |e^{g(z)-g(z')}| \leq e^{2B}.$$

Let z be arbitrary in $\mathbf{D}_{r_0}$ and let $z' = z_2$. Then

$$|h(z)| \leq \frac{1}{\alpha} e^{2B}$$

which proves that $\{h\}$ is bounded. Similarly one shows that

$$\alpha e^{-2B} \leq |h(z)|.$$

It follows that $\{h\}$ is a normal family, as desired.

Case 1.2. The log derivative of f_2/f_1 is not bounded on $\mathbf{D}_{r_0}$.

Then we consider

$$\frac{f_1}{f_3} + \frac{f_2}{f_3} + 1 = 0.$$

Since we are in Case 1, we have the lower and upper bounds of α and $1/\alpha$ for all quotients. In addition, under this Case 1.2,

log derivative of $f_2/f_1 = \dfrac{f_2'}{f_2} - \dfrac{f_1'}{f_1} = L_g$

where $\quad g = (g_1, g_2), \quad g_1 = f_1/f_3, \quad g_2 = f_2/f_3.$

By Theorem 1.8 there exists a subsequence $\mathscr{S}$ of $f \in \mathscr{F}$ such that $\{f_1/f_3, f_2/f_3\}$ is a normal family. Then $\{f_1/f_3, f_2/f_3\}_{f \in \mathscr{S}}$ are **C***-convergent. This proves the theorem.

VIII, §3. ESTIMATES OF BLOCH–CARTAN

We recommend that the reader continue immediately with the next section and refer to the present one only as needed. A weaker version of the first theorem was stated by Bloch, who recognized its central nature, and apparently was not able to prove it.

This is one of the points which Cartan filled in, and to which Cartan alluded in his introduction when he said that Bloch's results did not seem to be completely proved.

Theorem 3.1. *In a euclidean space let* $P_1, \ldots, P_n$ *be a finite sequence of points (which may not be distinct). Let* $c > 0$. *Then the set of points* Q *for which one has the inequality*

$$\prod_{i=1}^{n} d(Q, P_i) \leqq c^n$$

is contained in the union of at most n balls such that the sum of the radii is bounded by 2ec.

Proof. For simplicity of notation, we denote the euclidean distance $d(Q, P)$ by QP. We reformulate the theorem slightly.

Theorem 3.2. *Let* $P_1, \ldots, P_n$ *be a finite sequence of points. Let* $R > 0$. *Then*

$$\frac{1}{n} \sum_{i=1}^{n} \log(QP_i) > \log R - 1$$

for all points Q *except possibly for those lying in at most n balls such that the sum of the radii is equal to* $2R$.

Observe that Theorem 3.1 follows from Theorem 3.2 by taking $R = ec$.

Now for the proof of the theorem. Suppose there exists a ball B containing all the points $P_1, \ldots, P_n$ and B has radius R. Let B' be the ball with the same center and twice the radius. If $Q \notin B'$ then $QP_i > R$ for all i, so

$$\log(QP_i) > \log R,$$

whence

$$\frac{1}{n} \sum_{i=1}^{n} \log(QP_i) > \log R,$$

and the theorem follows.

Suppose there is no ball as above. Then we let:

k_1 = largest integer such that there exists a ball B_1 of radius $k_1 R/n$ containing k_1 among the points $P_1, \ldots, P_n$;

k_2 = largest integer such that there exists a ball B_2 of radius $k_2 R/n$ containing k_2 among the points remaining after deleting the points contained in B_1;

k_3 = largest integer such that there exists a ball B_3 of radius $k_3 R/n$ containing k_3 among the points remaining after deleting the points contained in $B_1 \cup B_2$.

And so on. We get a sequence of balls $B_1,\ldots,B_p$ with $p \leq n$. Then:

1. To each point P_i is associated exactly one of the balls $B_1,\ldots,B_p$.
2. If r_j is the radius of B_j then $r_1 + \cdots + r_p = R$.
3. Given $m \leq n$, if there exists a ball B of radius mR/n containing $m' \geq m$ points among $P_1,\ldots,P_n$ then one of these m' points lies in a ball $B_1,\ldots,B_p$ of radius $\geq mR/n$.

Each of these properties is immediate from the construction.

Let $B'_1,\ldots,B'_p$ be the balls of twice the radius and centered at the same place as $B_1,\ldots,B_p$ respectively. Then the sum of the radii of $B'_1,\ldots,B'_p$ is $2R$.

Let Q be any point such that $Q \notin B'_1 \cup \cdots \cup B'_p$. We shall prove that

$$\frac{1}{n} \sum_{i=1}^{n} \log(QP_i) > \log R - 1.$$

We first prove:

Let $m \leq n$. Let $B(Q, mR/n)$ be the ball of center Q and radius mR/n. Then $B(Q, mR/n)$ contains at most $m - 1$ of the points $P_1,\ldots,P_n$.

Proof. Let $P_i \in B(Q, mR/n)$ and also $P_i \in B_j$ for some j. Let O_j be the center of B_j. Then

$$2r_j \leq QO_j < \frac{mR}{n} + r_j.$$

The right inequality is due to the fact that the two balls intersect in P_i. The left inequality comes from the assumption that $Q \notin B'_j$. It follows that

$$r_j < \frac{mR}{n}.$$

Hence by property 3, we see that $B(Q, mR/n)$ cannot contain $\geq m$ points among $P_1,\ldots,P_n$, thus proving our assertion.

Take $m = 1$. We conclude that there is no point P_i in $B(Q, R/n)$; there is at most one point P_i in $B(Q, 2R/n)$; there are at most two points among the P_i in $B(Q, 3R/n)$; and so forth. Then the sum $\sum \log(QP_i)$ is bounded from below by the estimate which arises when for each

$m = 1,\ldots,n$ there is one point among $P_1,\ldots,P_n$ which is at distance mR/n from Q. Thus we obtain

$$\frac{1}{n}\sum_{i=1}^{n}\log(QP_i) \geqq \frac{1}{n}\left(\log\frac{R}{n} + \log\frac{2R}{n} + \cdots + \log\frac{nR}{n}\right)$$

$$> \int_0^1 \log(Rt)\,dt = \log R - 1.$$

This proves Theorem 3.2.

Remark 1. It was precisely Theorem 3.1 which Bloch stated without proof, and which Cartan emphasized needed to be proved. Starting from the ball of largest radius is Cartan's essential step in making the proof go through.

Remark 2. Cartan extends the theorem to a sum

$$\frac{1}{n}\sum_{i=1}^{n} f(QP_i),$$

where f is an appropriate positive function, and also to the case when instead of a discrete distribution of points $P_1,\ldots,P_n$ there is a continuous distribution. These extensions won't be needed for our applications, and the reader is referred to Cartan for them.

In the second part of this section, which will use Theorem 1.1, we consider sums having to do with canonical Blaschke products.

Theorem 3.3. *Let $0 < c < 1/4e$ (for convenience). Let $\{b\}$ be a finite family of complex numbers in the disc $\mathbf{D}_r$ for $r < 1$. (These numbers may occur with multiplicities.) Then for $z, w \in \mathbf{D}_s$ with $0 < s < r < 1$ we have*

$$\sum_b \log\left|\frac{r^2 - \bar{b}z}{r(z - b)}\right| < M \sum_b \log\left|\frac{r^2 - \bar{b}w}{r(w - b)}\right|, \quad \text{with} \quad M = \frac{4r^2}{(r - s)^2}\log\frac{1}{c}$$

except for a set of points z contained in a finite union of discs such that the sum of the radii is equal to $4erc$, and so is $< 4ec$.

Proof. Letting $z \mapsto z/r$ and $w \mapsto w/r$ and $b \mapsto b/r$ we reduce the statement to the case when $r = 1$ and the points $\{b\}$ lie in $\mathbf{D} = \mathbf{D}_1$. In this case, the theorem reads:

Let $c > 0$. Let $\{a\}$ be a finite family of complex numbers in $\mathbf{D}$. Let $z, w \in \mathbf{D}_s$ with $s < 1$. Then

$$\sum_a \log\left|\frac{1 - \bar{a}z}{z - a}\right| < M \sum_a \log\left|\frac{1 - \bar{a}w}{w - a}\right|,$$

where $M = \dfrac{4}{(1 - s)^2} \log \dfrac{1}{c}$, except for a set of points z lying in the union of a finite number of discs such that the sum of the radii is equal to $4ec$.

Proof. We start from the inequality

$$\left|\frac{|z| - |a|}{1 - |a||z|}\right| \leqq \left|\frac{z - a}{1 - \bar{a}z}\right| \leqq \frac{|z| + |a|}{1 + |a||z|}.$$

One sees this inequality geometrically as follows. Let

$$T_a \colon \mathbf{D} \to \mathbf{D}$$

be the automorphism $T_a(z) = (a - z)/(1 - \bar{a}z)$. Then T_a interchanges 0 and a. In particular, T_a maps the line through 0, a onto itself, and maps the circle with center 0 and radius $|z|$ onto a circle whose diameter lies on the line through 0, a. The above inequality expresses the fact that the distance from 0 to a point on this circle is bounded by the distances from 0 to the extremeties of this diameter. After a rotation one may assume that a, z are on the real axis, and then the desired inequality drops out.

To prove the theorem, we consider two cases.

Case 1. $|a| > \dfrac{1 + s}{2}$. Then

$$\log\left|\frac{1 - \bar{a}z}{z - a}\right| \leqq \log \frac{1 - |a||z|}{|a| - |z|} = \log\left[1 + (1 - |a|)\frac{1 + |z|}{|a| - |z|}\right]$$

$$< (1 - |a|)\frac{1 + |z|}{|a| - |z|}$$

$$< \frac{4}{1 - s}(1 - |a|).$$

On the other hand,

$$\log\left|\frac{w-a}{1-\bar{a}w}\right| \leq \log\frac{|w|+|a|}{1+|a||w|} = \log\left[1-(1-|a|)\frac{1-|w|}{1+|a||w|}\right]$$

$$< -(1-|a|)\frac{1-|w|}{1+|a||w|},$$

whence

$$\log\left|\frac{1-\bar{a}w}{w-a}\right| > (1-|a|)\frac{1-|w|}{1+|a||w|} > \frac{1-s}{2}(1-|a|).$$

We then get

$$\log\left|\frac{1-\bar{a}z}{z-a}\right| < \frac{8}{(1-s)^2}\log\left|\frac{1-\bar{a}w}{w-a}\right| < M\log\left|\frac{1-\bar{a}w}{w-a}\right|,$$

because from the assumption $4ec < 1$ we see that

$$\frac{8}{(1-s)^2} < \frac{4}{(1-s)^2}\log(1/c),$$

so we get the right estimate on the right for each term in the sum over a in Case 1, in other words, we get

(1) $$\sum_a \log\left|\frac{1-\bar{a}z}{z-a}\right| < M\sum_a \log\left|\frac{1-\bar{a}w}{w-a}\right|,$$

where the sum is taken over a such that $|a| > (1+s)/2$.

Note that we did not use Theorem 3.1 yet.

Case 2. $|a| \leq \dfrac{1+s}{2}$. Then we use

$$\left|\frac{w-a}{1-\bar{a}w}\right| \leq \frac{|w|+|a|}{1+|a||w|}.$$

The maximum of the right-hand side occurs at the end points, with

$$|w| = s \qquad \text{and} \qquad |a| = \frac{1+s}{2}.$$

Therefore

$$\log\left|\frac{1-\bar{a}w}{w-a}\right| > \frac{(1-s)^2}{4}.$$

Let n be the number of points a with $|a| \leq (1 + s)/2$. Then we conclude that

$$\sum \log \left| \frac{1 - \bar{a}w}{w - a} \right| > n \frac{(1 - s)^2}{4},$$

where the sum is taken over those a with $|a| \leq (1 + s)/2$. But again taking the sum over these a, we have

$$\sum \log \left| \frac{1 - \bar{a}z}{z - a} \right| < \sum \log \frac{2}{|z - a|} \qquad \text{(trivially)}$$

$$< n \log \frac{1}{c} \qquad \text{(by Theorem 3.1)},$$

except for a set of z contained in a union of discs such that the sum of the radii is $4ec$. Therefore in Case 2, summing over those a such that $|a| \leq (1 + s)/2$ we finally get

$$(2) \quad \sum \log \left| \frac{1 - \bar{a}z}{z - a} \right| < M \sum \log \left| \frac{1 - \bar{a}w}{w - a} \right| \qquad \text{with} \quad M = \frac{4}{(1 - s)^2} \log \frac{1}{c}.$$

But we have now obtained the desired estimate in each case. Combining (1) and (2) concludes the proof of the theorem.

We shall give some applications of Theorem 3.3 in the form which will be used in the next sections.

Let f be meromorphic on **D**. *For $r < 1$ and $|z| < r$, $z \neq$ pole of f, we define*

$$N_{z, f}(r) = \sum_{\substack{b \in \mathbf{D}_r \\ b \neq 0}} \log \left| \frac{r^2 - \bar{b}z}{r(z - b)} \right|,$$

where the sum is taken over the poles of f with their multiplicities. Thus by definition if $f(0) \neq 0$ or ∞ we have

$$N_{0, f}(r) = N_f(r) = N_f(r, \infty).$$

Recall that all the logs occurring in the above sum are ≥ 0. In particular, $N_{z, f}(r) \geq 0$.

We also define the **height with the parameter** z by

$$T_{z, f}(r) = m_f(r) + N_{z, f}(r).$$

Our definition of $N_{z,f}(r)$ also shows that:

The inequality in Theorem 3.3 can be expressed as

$$N_{z,f}(r) < M N_{w,f}(r),$$

whence

$$T_{z,f}(r) < M T_{w,f}(r),$$

where z, w, r, and M have the meaning in the statement of Theorem 3.3.

The real Poisson–Jensen formula (1.5) after Theorem 1.4 of Chapter VI yields:

Proposition 3.4 (Poisson–Jensen inequality).

$$\log|f(z)| \leq \frac{r+|z|}{r-|z|} m_f(r) - \frac{r-|z|}{r+|z|} m_{1/f}(r) - N_{z,1/f}(r) + N_{z,f}(r).$$

Then the standard inequalities still hold, namely for f, g meromorphic on **D**,

$$N_{z,fg}(r) \leq N_{z,f}(r) + N_{z,g}(r),$$

$$T_{z,fg}(r) \leq T_{z,f}(r) + T_{z,g}(r).$$

We now formulate three variations of a certain situation where we apply Theorem 3.3 with these definitions. In the next two theorems, we are given a priori

$$\boxed{\alpha, \gamma > 0 \quad \text{and} \quad 0 < s < r_0 < 1.}$$

A relation concerning points w in some open set U will be said to hold γ-**almost everywhere on** U or for γ-**almost all** $w \in U$ if it holds for all $w \in U$ except possibly for w contained in a finite number of discs such that the sum of the radii is equal to γ. The results which follow combine the Poisson–Jensen inequality with Theorem 3.3. Because of their technical appearance, I again suggest that the reader refer to them only as needed for the proofs in the next section.

Theorem 3.5. *There exists a constant $K > 0$ depending only on the above numbers α, γ, s, r_0 having the following property. Let f be holomorphic on **D**. Suppose that there exists some $z \in \mathbf{D}_s$ such that*

$$|f(z)| > \alpha.$$

Then for $r_0 < r < 1$, we have for γ-almost all $w \in \mathbf{D}_s$ the inequality

$$T_{w,1/f}(r) < K + Km_f(r),$$

and in particular

$$m_{1/f}(r) < K + Km_f(r).$$

Proof. By the Poisson–Jensen inequality and the fact that $N_{z,f}(r) = 0$ since f is holomorphic, we get

$$m_{1/f}(r) + \frac{r+|z|}{r-|z|}N_{z,1/f}(r) \leqq \left(\frac{r+|z|}{r-|z|}\right)^2 m_f(r) - \frac{r+|z|}{r-|z|}\log|f(z)|.$$

But

$$-\log|f(z)| \leqq \log^+ \frac{1}{|f(z)|}$$

and since $r > r_0$, $|z| < s$, we get

$$m_{1/f}(r) + N_{z,1/f}(r) \leqq \left(\frac{r_0+s}{r_0-s}\right)^2 m_f(r) + \frac{r_0+s}{r_0-s}\log^+\frac{1}{|f(z)|},$$

whence

$$T_{z,1/f}(r) \leqq \left(\frac{r_0+s}{r_0-s}\right)^2 m_f(r) + \frac{r_0+s}{r_0-s}\log^+ 1/\alpha.$$

Since $T_{w,1/f}(r) \leqq MT_{z,1/f}(r)$ by Theorem 3.3 we have concluded the present proof.

Remark. The last inequality in Theorem 3.5 does not depend on z and is already interesting. We can obtain it more shortly by selecting z such that $|f(z)| > \alpha$, and deleting the reference to $N_{z,1/f}$. For instance, we use this special case in the proof of Lemma 4.3 below.

Theorem 3.6. *Let f be meromorphic on $\mathbf{D}$. Assume that the set of points $z \in \mathbf{D}_s$ for which*

$$|f(z)| > \alpha$$

is not contained in a finite union of discs such that sum of the radii is equal to γ. Then:

(i) *For every $w \in \mathbf{D}_s$ and $r_0 < r < 1$, we have*

$$m_{1/f}(r) < K + KT_{w,f}(r).$$

(ii)   *We also have for $r_0 < r < 1$*

$$T_{w,\,1/f}(r) < K + K T_{w,\,f}(r)$$

for γ-almost all $w \in \mathbf{D}_s$.

Proof. By the Poisson–Jensen inequality and the fact that $N_{z,\,1/f} \geq 0$ we have

$$m_{1/f}(r) \leq \left(\frac{r_0 + s}{r_0 - s}\right)^2 T_{z,\,f}(r) + \frac{r_0 + s}{r_0 - s} \log^+ \frac{1}{|f(z)|}$$

for all z. By the hypothesis and Theorem 3.3 we can choose z such that both inequalities

$$T_{z,\,f}(r) \leq M T_{w,\,f}(r) \qquad \text{and} \qquad |f(z)| > \alpha.$$

hold. This proves (i).

For (ii), we write the Poisson–Jensen inequality in the form

$$T_{z,\,1/f}(r) \leq \left(\frac{r_0 + s}{r_0 - s}\right)^2 T_{z,\,f}(r) + \frac{r_0 + s}{r_0 - s} \log^+ \frac{1}{|f(z)|}.$$

Again we choose z such that both inequalities hold:

$$T_{z,\,f}(t) \leq M T_{w,\,f}(r) \qquad \text{and} \qquad |f(z)| > \alpha.$$

For all $w \in \mathbf{D}_s$ except for a set of points contained in a finite union of discs whose sum of radii is equal to γ, Theorem 3.3 gives the inequality

$$T_{w,\,1/f}(r) < M T_{z,\,1/f}(r),$$

which proves (ii), and concludes the proof of the theorem.

VIII, §4. CARTAN'S CONJECTURE AND THE CASE OF FOUR FUNCTIONS

We start with the conjecture because it tells us where things are going.

Cartan's conjecture. *Let $\mathcal{F}$ be an infinite family of p-tuples of units $f = (f_1, \ldots, f_p)$ on $\mathbf{D}$ satisfying Borel's equation*

$$f_1 + \cdots + f_p = 0.$$

*Then there exists a subsequence $\mathscr{S}$ having the following property.
There exists a partition of the indices $\{1,\ldots,p\}$ into disjoint sets $\{S\}$
and each S contains a subset I with at least two elements, which may
equal S itself. These satisfy the following properties for $f \in \mathscr{S}$:*

(i) *For each S and all $i,j \in I$ the sequence $\{f_i/f_j\}$ is $\mathbf{C}^*$-convergent.*

(ii) *If $k \in S$ and $k \notin I$ then $\{f_k/f_i\}$ converges to 0.*

(iii) *Finally given $i \in I$,*

$$\sum_{j \in I} f_j/f_i \quad converges\ to\ 0.$$

The conjecture has been proved when $p = 3$ and will be completely
proved when $p = 4$ but will be only partially proved for arbitrary p.
However, enough of the conjecture will be proved to yield Theorem 3.8
of Chapter III. Indeed, consider the imbedding of $\mathbf{P}^n$ into $\mathbf{P}^{n+1}$ as in
Chapter III, §3 so $\mathbf{P}^n$ is the hyperplane

$$x_0 + \cdots + x_{n+1} = 0.$$

Let H_i be the hyperplane $x_i = 0$. Then the diagonals are defined by the
equations

$$\sum_{i \in I} x_i = 0,$$

where I contains at least two indices and at most n.

Both in the case of three and four functions and the case of arbitrarily
many functions in the next section, the theorems imply that if we let

$$Y = \mathbf{P}^n, \qquad X = \mathbf{P}^n - \{\text{the } n + 2 \text{ hyperplanes}\},$$

$$\Delta = \text{union of the diagonals},$$

then $\mathrm{Hol}(\mathbf{D}, X)$ is relatively locally compact in $\mathrm{Hol}(\mathbf{D}, Y)$ mod Δ, and in
particular X is hyperbolically imbedded in Y mod Δ as stated in
Theorem 1.4 of Chapter II, thus proving Theorem 3.8 of Chapter III. Indeed, although Cartan did not succeed in proving that the entire set of
indices could be covered by sets S as in the conjecture, he at least shows
that either there is one such set S which is equal to all the indices, or
there are at least two such sets S. In the first case, there is a subsequence which converges uniformly on compact sets. In the second case,
a subsequence converges to a diagonal Δ_I for some I, which contains at
least two elements and not more than n when the set of indices is now
$\{0,\ldots,n+1\}$, because of the last condition

$$\sum_{j \in I} f_j/f_i \quad \text{convergent to } 0.$$

We now repeat the theorem we have already proved for three functions, since we shall carry out an induction, and we want the statement ready for use, uncluttered by other proofs or lemmas.

Theorem 2.1. *Suppose we have an infinite family $\mathscr{G}$ of systems of three units on $\mathbf{D}$ satisfying*

$$g_1 + g_2 + g_3 = 0 \qquad \text{with} \quad g \in \mathscr{G}.$$

Then there exists an infinite subsequence $\mathscr{S}$ satisfying one of the following conditions:

(a) *All quotients g_i/g_j are $\mathbf{C}^*$-convergent for $g \in \mathscr{S}$.*
(b) *There exists a subset $I = \{i, j\}$ with two elements such that g_i/g_j is $\mathbf{C}^*$-convergent, and if $k \notin I$ then both g_k/g_i and g_k/g_j converges to 0.*

In both cases **(a)** and **(b)** we say that the indices in I are of **first type**, and the complement of I is of **second type**. In **(a)**, I consists of all three indices. In **(b)**, I consists of two indices.

Remark. When we apply Theorem 2.1 via an inductive procedure, we shall in fact apply the corresponding theorem for domains of definitions of the functions which may be more general open sets U. Indeed, if a sequence of holomorphic functions has a subsequence which converges uniformly on some neighborhood of each point of U, then by a diagonal procedure, we can find a subsequence which converges uniformly on each compact subset of U. Or to put it another way, if a family of holomorphic functions is normal when restricted to some disc around each point of U, then the family itself is normal on U.

Next we state the theorem for four functions.

Theorem 4.1. *Let $\mathscr{F}$ be an infinite family of four units f on $\mathbf{D}$ which are solutions of*

$$f_1 + f_2 + f_3 + f_4 = 0.$$

Then there exists an infinite subsequence $\mathscr{S}$ satisfying one of the following conditions for $f \in \mathscr{S}$:

(a) **1.** *There is a subset of the indices I containing at least two elements such that all the quotients f_i/f_j are $\mathbf{C}^*$-convergent for $i, j \in I$.*
       2. *If $k \notin I$ then f_k/f_i converges to 0 for all $i \in I$.*

(b) *The set of indices* $\{1, 2, 3, 4\}$ *is the disjoint union of two subsets* I, J *each having two elements, and the quotients* f_i/f_j *converge to* -1 *for each pair* $i \neq j$ *in each subset.*

We observe that in case (a), given $i \in I$, it *follows* from what we stated when $n = 4$ that

$$\sum_{j \in I} f_j/f_i \quad \text{converges to } 0.$$

Also we note that in case (a), in Cartan's conjecture for $n = 4$, there is exactly one set S consisting of all indices. In case (b), there are two sets S_1 and S_2 which cover $\{1, 2, 3, 4\}$.

Indices in the sets I will be said to be of **first type**. Indices in the complementary sets $S - I$ will be said to be of **second type**, both in the conjecture and in Theorem 4.1.

In case (a) we say nothing about the quotients of functions of second type among themselves. Cartan gives the following example when such quotients do not converge. Let

$$f_1(z) = 1 - e^{n(z-2)},$$

$$f_2(z) = -1 - e^{2n(z-1)},$$

$$f_3(z) = e^{2n(z-1)},$$

$$f_4(z) = e^{n(z-2)},$$

where n is the sequence of positive integers. Then f_3/f_4 is not a normal family since f_3/f_4 converges to 0 if the real part of z is negative, and converges to ∞ if this real part is positive.

In case (b) we see that the set of indices decomposes into the union of two disjoint subsets, and that we say nothing about the quotients f_k/f_i with i in one subset and k in the complement. Indeed, a sequence of such functions need not converge. Cartan gives the following example:

$$f_1(z) = -e^{nz},$$

$$f_2(z) = e^{nz}(1 - e^{-n}),$$

$$f_3(z) = -1,$$

$$f_4(z) = 1 + e^{n(z-1)}.$$

The family f_1/f_3 is not normal. One could also take $f_2 = -f_1$ and $f_4 = -f_3$ with f_1/f_3 not normal.

The rest of this section is devoted to the proof of Theorem 4.1.

It suffices to prove the convergence properties on a disc $\mathbf{D}_{r_0}$. We then let

$$r_0 < r_1 < 1.$$

Case 0.1. There exists a subsequence $\mathscr{S}$ such that for $f \in \mathscr{S}$ and some pair of indices i, k, f_k/f_i converges to 0 on $\mathbf{D}_{r_1}$.

Say f_4/f_1 converges to 0. We consider the identity

$$\left(1 + \frac{f_4}{f_1}\right) + \frac{f_2}{f_1} + \frac{f_3}{f_1} = 0.$$

We let

$$g_1 = 1 + \frac{f_4}{f_1}, \qquad g_2 = \frac{f_2}{f_1}, \qquad g_3 = \frac{f_3}{f_1}.$$

After subsequencing if necessary, we may assume that g_1 has no zero on $\mathbf{D}_{r_0}$. Thus g_1, g_2, g_3 are units, and we can apply Theorem 2.1 which treats solutions of the Borel equation with three units. We look at the different cases in that theorem.

If all three indices 1, 2, 3 are of first type for the sequence $\mathscr{G} = \{g\}$, then we let $I = \{1, 2, 3\}$. Then case (a 1) is satisfied for this set. There is one index, namely 4, in the complementary set, and f_4/f_i converges to 0 for $i \in I$ by hypothesis, so (a 2) is also satisfied. This proves the theorem if all indices 1, 2, 3 are of first type.

Suppose that the index 1 is of second type for $\mathscr{G}$, and thus that 2, 3 are of first type. Then by assumption, both g_1/g_2 and g_1/g_3 converge to 0. Since g_1 converges to 1, it follows that f_1/f_2 and f_1/f_3 converges to 0, whence f_4/f_2 and f_4/f_3 converge to 0. Then finally f_3/f_2 converges to -1. Then we let $I = \{2, 3\}$ and see that condition (a) is satisfied.

Suppose on the other hand that 2 or 3, say 3, is of second type for $\mathscr{G}$. Then f_3/f_1 and f_4/f_1 converge to 0, whence f_2/f_1 converges to -1. Again we have condition (a).

This proves the theorem in Case 0.1.

Case 0.2. There exists a subsequence $\mathscr{S}$ such that for some pair of indices i, j and $f \in \mathscr{S}$ the quotient f_i/f_j is $\mathbf{C}^*$-convergent on $\mathbf{D}_{r_1}$ to a limit $\neq -1$.

Say f_4/f_1 is $\mathbf{C}^*$-convergent to a limit $\neq -1$. Let g_1, g_2, g_3 be as in the preceding case. Then g_1 is not the zero function, and has only a finite number of zeros in $\mathbf{D}_{r_0}$. Let U be the open subset of $\mathbf{D}_{r_0}$ obtained by deleting small discs centered at the zeros of g_1. Subsequencing if necessary, we may assume that g_1 has no zeros on U. We can then argue as in the preceding Case 0.1, taking into account the remark that Theorem 2.1 applies to an arbitrary open set U instead of $\mathbf{D}$. We then conclude that the various convergences of the theorem occur on U. But if $\{h_n\}$ is a sequence of holomorphic functions on a closed disc (i.e. an open set containing a closed disc), and $\{h_n\}$ converges uniformly on the boundary, then Cauchy's formula shows that $\{h_n\}$ converges uniformly on compact subsets of the disc. Then the desired convergences for f_1, f_2, f_3, f_4 occur on all of $\mathbf{D}_{r_0}$, not just U. For instance, suppose f_i/f_j is $\mathbf{C}^*$-convergent on U. Since f_i/f_j and its inverse f_j/f_i are both holomorphic, they are $\mathbf{C}^*$-convergent on the interiors of the above discs. This concludes the present case.

The next cases are the ones which will involve Wronskians, and we return to prove a lemma about them.

Let $f_1, \dots, f_n$ be meromorphic functions. We follow the Bloch–Cartan notation and for simplicity write

$$W(f_1 \cdots f_n) = |f_1 \cdots f_n| = \begin{vmatrix} f_1 & \cdots & f_n \\ \vdots & & \vdots \\ f_1^{(n-1)} & \cdots & f_n^{(n-1)} \end{vmatrix}.$$

We note that for any functions g_1, g_2,

$$L(g_1, g_2) = \frac{|g_1 g_2|}{g_1 g_2} = \frac{g_2'}{g_2} - \frac{g_1'}{g_1}$$

is the logarithmic derivative of g_2/g_1.

Lemma 4.2. *Let*

$$g_1 = |f_1 f_3 \cdots f_n| \qquad \textit{and} \qquad g_2 = |f_2 f_3 \cdots f_n|.$$

Then

$$|g_1 g_2| = |f_3 \cdots f_n| |f_1 f_2 \cdots f_n|$$

Proof. What we are really dealing with is a differential field, i.e. a field with a derivation. It suffices to prove the relation in the generic case. Namely, let $x_1, \dots, x_n, x_1', \dots, x_n', \dots, x_1^{(n-1)}, \dots, x_n^{(n-1)}$ be n^2 independent variables. We define

$$|x_1 \cdots x_n|, \qquad y_1 = |x_1 x_3 \cdots x_n|, \qquad y_2 = |x_2 x_3 \cdots x_n|$$

by the similar determinants as for functions. It suffices to prove the desired relation in the case of these variables, that is

$$|y_1 y_2| = |x_3 \cdots x_n| \, |x_1 x_2 \cdots x_n|.$$

Let (a_i^j) be n^2 numbers satisfying the polynomial equation

$$|x_1 \cdots x_n| = 0.$$

It will suffice to prove that they also satisfy the equation

$$|y_1 y_2| = 0.$$

Indeed, this will imply that $|x_1 \cdots x_n|$ divides $|y_1 y_2|$. Equating the coefficients of $x_1^{(n-1)}$ gives the desired identity.

So suppose

$$\begin{vmatrix} a_1^0 & a_2^0 & \cdots & a_n^0 \\ a_1^1 & a_2^1 & \cdots & a_n^1 \\ \vdots & & & \vdots \\ a_1^{(n-1)} & a_2^{(n-1)} & \cdots & a_n^{(n-1)} \end{vmatrix} = 0.$$

The n functions

$$P_i(t) = a_i^0 + a_i^1 t + \cdots + \frac{1}{(n-1)!} a_i^{n-1} t^{n-1} \qquad (i = 1, \ldots, n)$$

satisfy a linear relation

$$\alpha_1 P_1 + \cdots + \alpha_n P_n = 0$$

with not all coefficients equal to 0. If $\alpha_1 = \alpha_2 = 0$ then $P_3, \ldots, P_n$ are linearly dependent, so y_1, y_2 vanish if we substitute $P_1, \ldots, P_n$ for $x_1, \ldots, x_n$ respectively, whence $|y_1 y_2| = 0$ under the substitution. On the other hand, if α_1, α_2 are not both 0, then the $n - 1$ functions

$$\alpha_1 P_1 + \alpha_2 P_2, P_3, \ldots, P_n$$

are linearly dependent, so

$$\alpha_1 y_1 + \alpha_2 y_2 = 0$$

under this same substitution, whence also

$$|y_1 y_2| = 0$$

under this substitution. Finally we let $t = 0$ in this last identity to conclude the proof of the lemma.

The lemma will be used just as we used logarithmic derivatives in the case of three functions, but in a more complicated way, which will use Theorem 3.5.

By Lemma 4.2. we see that

$$\text{log derivative of } \frac{|f_2 f_3 \cdots f_n|}{|f_1 f_3 \cdots f_n|} = \frac{|f_3 \cdots f_n|}{|f_1 f_3 \cdots f_n|} \frac{|f_1 f_2 \cdots f_n|}{|f_2 f_3 \cdots f_n|}.$$

Such a logarithmic derivative will be called an **n-term derived fraction**. Thus a **two-term derived fraction** is an expression like

$$\frac{|f_1 f_2|}{f_1 f_2} = \text{log derivative of } f_2/f_1.$$

A family $\{f_1, \ldots, f_n\}$ of n functions gives rise to $n(n-1)/2$ n-term derived fractions, which come in pairs up to a sign, depending on the choice of the two functions playing the special role of f_1, f_2 above. So for $n = 4$, we get six 4-term derived fractions.

We observe that the n-term derived fractions are homogeneous of degree 0 in $(f_1, \ldots, f_n)$ by Proposition 1.5 of Chapter VII.

A quotient like

$$Q = \frac{|f_1 f_2|}{|f_1 f_3|}$$

can be expressed as a product

$$\frac{|f_1 f_2|}{|f_1 f_3|} = \frac{|f_1 f_2|}{f_1 f_2} \frac{f_1 f_3}{|f_1 f_3|} \frac{f_2}{f_3}.$$

Thus

$Q'/Q = \text{sum of 3-term derived fractions} + \text{2-term derived fraction}.$

We shall use the identity to estimate Q'/Q, but also to express f_2/f_3, say, in terms of derived fractions, and expressions like Q to estimate such quotients f_2/f_3.

Case 1. *We suppose that Case* 0.1 *is not satisfied. Then there exists some* $\alpha > 0$ *having the following property. For each quotient* f_i/f_j *there exists* z_1, $z_2 \in \mathbf{D}_{r_1}$ *such that*

$$\alpha < |f_i/f_j(z_1)| \quad and \quad |f_i/f_j(z_2)| < 1/\alpha.$$

We meet subcases depending on the 2-term derived fractions.

Case 1.1. *There exists a subsequence* $\mathscr{S}$ *such that two of the 2-term derived fractions are bounded by* 1 *in absolute value on* $\mathbf{D}_{r_1}$.

Suppose for instance that $|f_1 f_2|/f_1 f_2$ is such a derived fraction. Then by Case 1.1 of Theorem 2.1, especially Lemma 2.2 used in that case, after subsequencing we may assume that f_2/f_1 is $\mathbf{C}^*$-convergent on $\mathbf{D}_{r_1}$. If the limit of f_2/f_1 is $\neq -1$ then we are in Case 0.2 and the theorem is proved for $\mathbf{D}_{r_0}$.

If f_2/f_1 is $\mathbf{C}^*$-convergent to -1, then we look at the second 2-term derived fraction, and conclude the proof similarly unless the limit is again -1.

Suppose this second derived fraction is $|f_i f_j|/f_i f_j$ and the limit of f_j/f_i is -1. If i or j is equal to 1 or 2, say $i = 1$, $j = 3$, then f_3/f_1 converges to -1, f_3/f_2 converges to 1, and we are now in Case 0.2, which concludes the proof.

On the other hand, if, say, $i = 3$ and $j = 4$, then f_3/f_4 converges to -1. Then Condition **(b)** of Theorem 4.1 prevails, and so Theorem 4.1 is proved in Case 1.1.

We now suppose that Case 1.1 *is not satisfied. Then there exists a subsequence* $\mathscr{S}$ *having the following property. If* $f \in \mathscr{S}$ *and* $L_2 = |f_i f_j|/f_i f_j$ *denotes five out of six of the 2-term derived fractions, then for each such* L_2 *there exists* $z \in \mathbf{D}_{r_1}$ *such that* $|L_2(z)| \geq 1$.

To fix ideas, we suppose that the possibly exceptional 2-term derived fraction is $|f_3 f_4|/f_3 f_4$.

Recall that from Lemma 4.2 we know that

$$\text{log derivative of } \frac{|f_1 f_2|}{|f_1 f_3|} = \frac{-f_1 |f_1 f_2 f_3|}{|f_1 f_2| |f_1 f_3|}.$$

We then come to the next case.

Case 1.2. *Let* $r_1 < r_2' < r_2$. *There exists a subsequence* $\mathscr{S}'$ *of* $\mathscr{S}$ *having the following property. Let*

$$\gamma = \frac{r_2 - r_2'}{12}.$$

Let $f \in \mathscr{S}'$. Let L_3 denote any one of the 3-term derived fractions formed with f_1, f_2, f_3. Then for γ-almost all $z \in \mathbf{D}_{r_2}$ we have

$$|L_3(z)| \leqq 1.$$

For concreteness we note that L_3 can be any one of the three functions

$$\frac{f_1 |f_1 f_2 f_3|}{|f_1 f_2| \, |f_1 f_3|}, \qquad \frac{f_2 |f_1 f_2 f_3|}{|f_2 f_1| \, |f_2 f_3|}, \qquad \frac{f_3 |f_1 f_2 f_3|}{|f_3 f_1| \, |f_3 f_2|}.$$

We shall prove:

In Case 1.2, there exists a subsequence $\mathscr{S}''$ of $\mathscr{S}'$ such that for all $f \in \mathscr{S}''$ all quotients $f_i/f_j \, (i, j = 1, 2, 3)$ are $\mathbf{C}^*$-convergent on $\mathbf{D}_{r_1}$, and so Case 1.2 implies Case 0.2.

Indeed not all three such quotients can converge to -1, so it follows that we are now in Case 0.2 which has already been treated, and therefore the theorem will be proved in this case.

The proof is based on the following Lemma.

Lemma 4.3. *Suppose we are in Case* 1.2. *Write* $m(r, h)$ *for* $m_h(r)$ *to avoid complicated indices. Define*

$$m(r) = \max m(r, f_i/f_j) \qquad \textit{for} \quad i \neq j, \qquad i, j = 1, 2, 3.$$

Let $L_{ij} = |f_i f_j|/f_i f_j$ *for these same* i, j. *Then for* $f \in \mathscr{S}'$:

$$m(r) \leqq K \max m(r, L_{ij})$$

for all r *in disjoint intervals lying in* $[r'_2, r_2]$ *such that the sum of the lengths of these intervals is at least* $(r_2 - r'_2)/2$. *The letter* K *denotes a constant independent of all* $f \in \mathscr{S}'$ *and of* r.

Proof. The proof of this lemma will occur in two steps.

Step 1. We consider the band

$$r'_2 < |z| < r_2.$$

We shall now delete from this band a finite number of smaller bands where the inequality of Case 1.2 does not hold for a given $f \in \mathscr{S}'$. By hypothesis, there is a finite number of discs such that the sum of the radii is 3γ, and such that if z does not lie in these discs, then $|L_3(z)| \leqq 1$ for any 3-term derived fraction L_3 formed with f_1, f_2, f_3. For each disc, we delete the band centered at the origin, and formed with the two circles tangent to this disc.

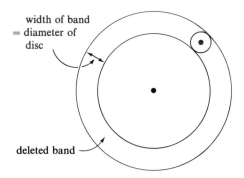

width of band
= diameter of
disc

deleted band

Then we have deleted a finite number of bands, centered at the origin, such that the sum of their widths is at most 6γ. By our choice of γ, the complement of the deleted bands in $r'_2 < |z| < r_2$ consists of a finite number of bands $B_1, \ldots, B_m$ such that the sum of their width is at least $(r_2 - r'_2)/2$. Let $z \in B_k$ for $k = 1, \ldots, m$. Then

$$|L_3(z)| \leq 1.$$

We now prove:

Let Q be any one of the quotients

$$Q = \frac{|f_1 f_2|}{|f_1 f_3|} \quad or \quad \frac{|f_2 f_1|}{|f_2 f_3|} \quad or \quad \frac{|f_3 f_1|}{|f_3 f_2|}.$$

Then for z, $z' \in B_k$ the quotient $|Q(z)/Q(z')|$ is bounded above and below away from 0 by positive constants depending only on r_2, r'_2.

Proof. By assumption, $|Q'/Q(z)| \leq 1$ for all z in a band B_k. We can join z to z' by a curve whose length is bounded by some fixed number A (which could be taken to be $\pi r_2 + r_2 - r'_2$ for instance). Integrating along this curve, choosing a fixed determination of $\log Q(z)$, and getting $\log Q(z')$ by continuation along the curve yields

$$|\log Q(z) - \log Q(z')| \leq A.$$

Since for complex u we have $-|\log u| \leq \log|u| \leq |\log u|$, we get

$$-A \leq \log|Q(z)/Q(z')| \leq A,$$

thus proving our assertion after taking exp.

As a consequence, we see that if $|Q(z)| \geqq 1$ for some point z of a band B_k, then $|Q|$ is bounded from below by e^{-A} on this band. Otherwise, $|Q|$ is bounded from above by e^A.

Step 2. Let r be the radius of a circle centered at the origin and lying in the interior of one of the bands $B_1, \ldots, B_m$ of step 1. Then

$$m(r, f_2/f_3) \quad and \quad m(r, f_3/f_2) \leqq K + Km\left(r, \frac{|f_1 f_2|}{f_1 f_2}\right) + Km\left(r, \frac{|f_1 f_3|}{f_1 f_3}\right).$$

Note that Lemma 4.3 follows at once from Step 2, which we now proceed to prove.

Let $Q = |f_1 f_2|/|f_1 f_3|$. Suppose that $|Q|$ is bounded from above on the circle $|z| = r$ for $f \in \mathscr{S}'$, as at the end of Step 1. The case when $|Q|$ is bounded from below would be treated the same way. Write

$$\frac{f_2}{f_3} = \frac{|f_1 f_2|}{|f_1 f_3|} \cdot \frac{|f_1 f_3|}{f_1 f_3} \cdot \frac{f_1 f_2}{|f_1 f_2|}.$$

Then by the assumption on Q we get

$$m(r, f_2/f_3) < K + m\left(r, \frac{|f_1 f_3|}{f_1 f_3}\right) + m\left(r, \frac{f_1 f_2}{|f_1 f_2|}\right)$$

$$< K + m\left(r, \frac{|f_1 f_3|}{f_1 f_3}\right) + Km\left(r, \frac{|f_1 f_2|}{f_1 f_2}\right) \qquad \text{by Theorem 3.5.}$$

But since

$$|f_3/f_2(z_1)| < \alpha \qquad \text{for some} \quad z_1 \in \mathbf{D}_{r_1},$$

again by Theorem 3.5 and the trivial inequality $m_h \leqq T_{w,h}$ we get

$$m(r, f_3/f_2) < K + Km(r, f_2/f_3).$$

This concludes the proof of Step 2, and hence the proof of Lemma 4.3.

Now we show how Lemma 4.3 implies the conclusion of Case 1.2. We can apply the method of §1. The inequality of Lemma 4.3 is analogous to the inequalities (*) and (**) of Lemma 1.3. The same argument as in the proof of Lemma 1.3 leads to the inequality as in that lemma. One can continue as in Lemmas 1.4 and 1.5, except that instead of having intervals such that the sum of the lengths is bounded by

$$2e^{-S(r)/\lambda} < 1 - r,$$

we now have intervals the sum of whose lengths is $< (r_2 - r_2')/2$. Then finally we find that

$$m(r) \leqq \text{maximum of } M, 2B \log \frac{4}{r_2 - r_2'},$$

independently of $f \in \mathscr{S}'$. In particular, $m(r_1)$ is bounded, and Montel's Theorem 1.6 applies to conclude the proof.

Next we show:

If Case 1.2 is not satisfied, then there exists a subsequence $\mathscr{S}'$ such that for all $f \in \mathscr{S}'$ all quotients f_i/f_j ($i,j = 1, 2, 3, 4$) are $\mathbf{C}^*$-convergent on $\mathbf{D}$.

This will of course conclude the proof of Theorem 4.1.

By definition if Case 1.2 is not satisfied, there exists a subsequence $\mathscr{S}'$ and a 3-term derived fraction, say L_3, such that for $f \in \mathscr{S}'$, the set of points $z \in \mathbf{D}_{r_2}$ such that $|L_3(z)| \geq 1$ cannot be covered by a finite number of discs whose sum of the radii is equal to γ. For concreteness, we let

$$L_3 = \frac{f_3 |f_1 f_2 f_3|}{|f_3 f_1| |f_3 f_2|} = L_3(f), \qquad f \in \mathscr{S}'.$$

We then rely on an analogue of Lemma 4.3.

Lemma 4.4. *Suppose we are not in Case 1.2. Let $r_2 < r_3 < 1$. Then for $f \in \mathscr{S}'$ and $r_3 < r < 1$ we have*

$$m\left(r, \frac{f_1 f_2 f_3}{|f_1 f_2 f_3|}\right) < K + Km\left(r, \frac{|f_1 f_2 f_3|}{f_1 f_2 f_3}\right)$$
$$+ Km\left(r, \frac{|f_1 f_3|}{f_1 f_3}\right) + Km\left(r, \frac{|f_2 f_3|}{f_2 f_3}\right),$$

where K is a constant independent of $f \in \mathscr{S}'$ and of r.

Proof. As in Bloch, we have the identity

(∗) $$\frac{|f_1 f_2 f_3|}{f_1 f_2 f_3} = \frac{f_3 |f_1 f_2 f_3|}{|f_1 f_3| |f_2 f_3|} \frac{|f_1 f_3|}{f_1 f_3} \frac{|f_2 f_3|}{f_2 f_3},$$

whence

$$m\left(r, \frac{f_1 f_2 f_3}{|f_1 f_2 f_3|}\right)$$

$$\leqq m\left(r, \frac{|f_1 f_3| |f_2 f_3|}{f_3 |f_1 f_2 f_3|}\right) + m\left(r, \frac{f_1 f_3}{|f_1 f_3|}\right) + m\left(r, \frac{f_2 f_3}{|f_2 f_3|}\right)$$

$$\leqq K + KT_z\left(r, \frac{f_3 |f_1 f_2 f_3|}{|f_1 f_3| |f_2 f_3|}\right) + Km\left(r, \frac{|f_1 f_3|}{f_1 f_3}\right) + Km\left(r, \frac{|f_2 f_3|}{f_2 f_3}\right)$$

for all $z \in \mathbf{D}_{r_1}$ by Theorem 3.6(i) for the first term, and by Theorem 3.5 for the other two terms on the right since we have supposed that Case 1.1 is not satisfied. Furthermore, by using (∗) in a form expressing

$$f_3 |f_1 f_2 f_3| / |f_1 f_3| |f_2 f_3|$$

in terms of the other products, and the submultiplicativity of T_z together with the fact that $T_z = m_z$ for a holomorphic function, we find

$$T_z\left(r, \frac{f_3 |f_1 f_2 f_3|}{|f_1 f_3| |f_2 f_3|}\right) \leqq m_z\left(r, \frac{|f_1 f_2 f_3|}{f_1 f_2 f_3}\right) + T_z\left(r, \frac{f_1 f_3}{|f_1 f_3|}\right) + T_z\left(r, \frac{f_2 f_3}{|f_2 f_3|}\right).$$

Now the last two terms on the right with T_z can be estimated by Theorem 3.5 just as in the previous step to conclude the proof of the lemma.

This brings us to the situation of Theorem 1.9. We let

$$g_i = -f_i/f_4 \quad \text{for} \quad i = 1, 2, 3 \quad \text{so that} \quad g_1 + g_2 + g_3 = 1.$$

From this we conclude that $\{g_1\}$, $\{g_2\}$, $\{g_3\}$ are $\mathbf{C}^*$-convergent on $\mathbf{D}$. This finishes the proof when Case 1.2 is not satisfied, and also finishes the proof of Theorem 4.1.

For the inductive step, we need Theorem 4.1 not just on the unit disc but on an arbitrary open set U, in other words:

Theorem 4.5. *Theorem* 4.1 *remains valid if instead of a family of units on* $\mathbf{D}$ *we have a family of units on an arbitrary open set* U.

Proof. As Cartan remarks, one cannot argue naively as for the case of three functions, because it is not necessarily true any more that if Theorem 4.1 is true on some disc centered at every point of a connected open set U, then it remains true on U. Indeed, the conclusion of

Theorem 4.1 does not allow us to conclude that the family $\{f_2/f_1\}$, for instance, is normal. Hence one has to argue differently, as follows.

Let U be open and connected. Then the universal covering space is the disc, and we may therefore view the family $\mathscr{F}$ as a family of functions on the disc. Then we can indeed apply Theorem 4.1 to conclude the proof in general.

VIII, §5. THE CASE OF ARBITRARILY MANY FUNCTIONS

Theorem 5.1. *Let $\mathscr{F}$ be an infinite family of p units on $\mathbf{D}$ satisfying*

$$f_1 + \cdots + f_p = 0.$$

Then there exists a subsequence $\mathscr{S}$ having one of the following properties.

(a) *The set of indices $\{1,\ldots,p\}$ satisfies Cartan's conjecture with a single set $S = \{1,\ldots,p\}$.*

(b) *There are two disjoint subsets S_1, S_2, each containing at least two elements, satisfying the three conditions (i), (ii), (iii) in Cartan's conjecture.*

We now see that the theorem indeed implies Theorem 3.8 of Chapter III, but fails to prove all of Cartan's conjecture because the subsets S are not shown to cover the whole set of indices $\{1,\ldots,p\}$. Furthermore, we see that Theorem 4.1 is a special case of Theorem 5.1, because when $p = 4$, the two subsets S_1, S_2 do cover $\{1, 2, 3, 4\}$.

Remark 5.2. *If Theorem 5.1 is true for units on $\mathbf{D}$ then it is also true for units on an arbitrary connected open set U.*

Proof. This is the same proof with the universal covering space $\mathbf{D}$ of U as the proof of Theorem 4.5.

The proof of Theorem 5.1 is by induction and will be outlined as in Cartan, since it involves no new technique except for the general formulation of the induction.

As in the cases with $p = 3$ and $p = 4$ it suffices to prove the convergence on a disc $\mathbf{D}_{r_0}$ with $r_0 < 1$. We let

$$r_0 < r_1 < 1.$$

We distinguish Case 0 and Case 1 as before, namely:

Case 0.1. *There exists a subsequence* $\mathscr{S}$ *such that for some pair of indices* k, i *the sequence* $\{f_k/f_i\}$ *with* $f \in \mathscr{S}$ *converges to* 0 *on* $\mathbf{D}_{r_1}$.

Case 0.2. *There exists a subsequence* $\mathscr{S}$ *such that for some pair of indices* $i \neq j$ *the sequence* $\{f_i/f_j\}$ *with* $f \in \mathscr{S}$ *is* **C***-convergent on* $\mathbf{D}_{r_1}$ *to a limit* $\neq -1$.

In both these cases, one can argue just as in Cases 0.1 and 0.2 when $p = 4$, and use induction, to conclude the proof of the theorem.

We then suppose that Case 0.1 is not satisfied. Then we get into Case 2, with $p - 2$ subcases as follows.

Case 1.1. *There exists a subsequence* $\mathscr{S}$ *such that for* $f \in \mathscr{S}$ *and all* 2-*term derived fractions* L_2 *we have*

$$|L_2(z)| \leq 1 \qquad \textit{for all} \quad z \in \mathbf{D}_{r_1}.$$

The same proof as in Theorem 4.1 applies to this case.

Suppose that Case 1.1 *is not satisfied. Then there exists a subsequence* $\mathscr{S}_1$ *such that if* L_2 *denotes any one of the* 2-*term derived fractions with only one possible exception, say* $|f_{p-1}f_p|/f_{p-1}f_p$, *we have*

$$|L_2(z_1)| \geq 1 \qquad \textit{for some} \quad z_1 \in \mathbf{D}_{r_1}.$$

From now on we consider only subsequences of $\mathscr{S}_1$. We get to the next case:

Case 1.2. *Let* $r_1 < r_2 < 1$. *There exists a subsequence* $\mathscr{S}'_1$ *and three indices* i, j, k *among* $\{1,\ldots,p-1\}$ *such that each* 3-*term derived fraction* L_3 *formed with* f_i, f_j, f_k *satisfies*

$$|L_3(z)| \leq 1$$

for γ_2-*almost all* $z \in \mathbf{D}_{r_2}$, *where* γ_2 *is suitably chosen.*

The same proof as in Theorem 4.1 applies to this case.

Suppose that Case 1.2 *is not satisfied. There exists a subsequence* $\mathscr{S}_2$ *of* $\mathscr{S}_1$ *having the following property. Let* i, j, k *denote three indices among* $\{1,\ldots,p\}$. *Then one of the three* 3-*term derived fractions* L_3 *formed with* f_i, f_j, f_k *satisfies*

$$|L_3(z)| \geq 1$$

for a set of points $z \in \mathbf{D}_{r_2}$ which is not contained in a finite number of discs whose sum of radii is equal to γ_2.

We then reach the next case:

Case 1.3. Let $r_2 < r_3 < 1$. There exists a subsequence $\mathscr{S}'_2$ of $\mathscr{S}_2$, and four indices i, j, k, l among $\{1, \ldots, p-1\}$, such that each one of the 4-term derived fractions L_4 formed with f_i, f_j, f_k, f_l satisfies

$$|L_4(z)| \leqq 1$$

for γ_3-almost all $z \in \mathbf{D}_{r_3}$.

In this case, one uses an identity of the form

$$\frac{f_k}{f_l} = \frac{|f_i f_j f_k|}{|f_i f_j f_l|} \frac{|f_i f_j f_l|}{f_i f_j f_l} \frac{f_i f_j f_k}{|f_i f_j f_k|}.$$

One gets a bound for

$$m\left(r, \frac{f_i f_j f_k}{|f_i f_j f_k|}\right)$$

as in Lemma 4.4, which leads to a proof of the theorem on the disc $\mathbf{D}_{r_0}$.

Thus the inductive pattern is established: at each step, we exclude the preceding case, and get into a new one, thus finding Cases 1.1, 1.2, ..., 1.$(p-2)$. The last step can be formulated formally as follows.

Case 1.$(p-2)$. Let $r_{p-3} < r_{p-2} < 1$. There exists a subsequence $\mathscr{S}'_{p-3}$ of $\mathscr{S}_{p-3}$ such that every $(p-1)$-term derived fraction L_{p-1} formed with $f_1, \ldots, f_{p-1}$ satisfies

$$|L_{p-1}(z)| \leqq 1$$

for γ_{p-2}-almost all $z \in \mathbf{D}_{r_{p-2}}$.

One proves the theorem in this case. Then finally:

Suppose that Case 1.$(p-2)$ is not satisfied. Then there exists a subsequence $\mathscr{S}_{p-2}$ of $\mathscr{S}_{p-3}$ for which there is a bound of

$$m\left(r, \frac{f_1 \cdots f_{p-1}}{|f_1 \cdots f_{p-1}|}\right)$$

as in Lemma 4.4, using Theorems 3.5 and 3.6.

In this final step, we can then apply Theorem 1.9 to the identity

$$\frac{f_1}{f_p} + \cdots + \frac{f_{p-1}}{f_p} + 1 = 0$$

to conclude the proof.

Remark. It seems likely that Cartan's "derived fractions" amount to the "derived curves" considered by Ahlfors, and that one could give a unified presentation of Cartan's work and that of Ahlfors. There are some obvious similarities between the two works, and unknown to Ahlfors, Cartan apparently anticipated the techniques, and went further in some respects.

Bibliography

[Ah]   L. AHLFORS, *The theory of meromorphic curves*, Acta Soc. Sci. Finn. N.S., A, **III** (1941), pp. 1–31.

[At]   M. ATIYAH, *Vector bundles over an elliptic curve*, Proc. London Math. Soc., **VII** (1957), pp. 414–452.

[Ax]   J. AX, *Some topics in differential algebraic geometry* II: *on the zeros of theta functions*, Amer. J. Math., **94**, (1972), pp. 1205–1213.

[Ba]   T. BARTH, *The Kobayashi distance induces the standard topology*, Proc. Amer. Math. Soc., **35**, No. 2 (1972), pp. 439–441.

[Bar]   C. BARTON, *Tensor products of ample vector bundles in characteristic p*, Amer. J. Math., **93** (1971), pp. 429–438.

[Bl]   A. BLOCH, *Sur les systèmes de fonctions holomorphes a variétés linéaires lacunaires*, Ann. École Normale, **43** (1926), pp. 309–362.

[Bo]   E. BOREL, *Sur les zéros des fonctions entières*, Acta Math., **20** (1887), pp. 357–396.

[Br]   R. BRODY, *Compact manifolds and hyperbolicity*, Trans. Amer. Math. Soc., **235** (1978), pp. 213–219.

[B–G]   R. BRODY and M. GREEN, *A family of smooth hyperbolic hypersurfaces in* P_3, Duke Math. J., **44** (1977), pp. 873–874.

[Ca–G] J. CARLSON and P. GRIFFITHS, *A defect relation for equidimensional holomorphic mappings between algebraic varieties*, Ann. Math., **95** (1972), pp. 557–584.

[Ca 1] H. CARTAN, *Sur les systèmes de fonctions holomorphes à variétés linéaires lacunaires et leurs applications*, Ann. École Normale, **45** (1928), pp. 255–346.

[Ca 2] H. CARTAN, *Sur les zéros des combinations linéaires de p fonctions holomorphes données*, Mathematica, **7** (1933), pp. 5–31. (The result was announced in C.R. Acad. Sci., **189** (1929), pp. 521–523, same title.)

[Ch]   S. CHERN, *On holomorphic mappings of Hermitian manifolds of the same dimension*, Proc. Symp. Pure Math., **11** (1968), *Entire Functions and Related Parts of Analysis*, Amer. Math. Soc., pp. 157–170.

[Co-G] M. Cowen and P. Griffiths, *Holomorphic curves and metrics of nega-tive curvature*, J. Analyse Math., **29** (1976), pp. 93–153

[Di 1] A. Dinghas, *Ein n-dimensionales Analogon des Schwarzes-Pickschen Flächensatzes fur holomorphe Abbildungen der komplexen Einheitskugel in ein Kähler-Mannigfaltigkeit*, Arb. für Forschung des Landes Nordrhein-Westfalen, **33** (1965), pp. 477-494.

[Di 2] A. Dinghas, *Uber das Schwarzsche Lemma und verwandte Satze*, Israel J. Math., **5** (1967), pp. 157–169.

[Du]   J. Dufresnoy, *Théorie nouvelle des familles complexes normales*, Ann. École Normale, **61** (1944), pp. 1–44.

[Fi]   G. Fischer, *Complex analytic geometry*, Springer Lecture Notes, Vol. 538, Springer-Verlag, Berlin, 1976.

[Fr]   K. Fritzsche, *q-konvexe Restmengen in kompakten komplexen Mannig-faltigkeiten*, Math. Ann., **221** (1976), pp. 251-273.

[Fu 1] H. Fujimoto, *On holomorphic maps into a taut complex space*, Nagoya Math. J., 46 (1972), pp. 49-61.

[Fu 2] H. Fujimoto, *Families of holomorphic maps into the projective space omitting some hyperplanes*, J. Math. Soc. Japan, **25**, No. 2 (1973), pp. 235-249.

[Fu 3] H. Fujimoto, *On meromorphic maps into the complex projective space*, J. Math. Soc. Japan, **26**, No. 2 (1974), pp. 272-288.

[Fu 4] H. Fujimoto, *Extensions of the big Picard theorem*, Tohoku Math. J., **24**, No. 3 (1972), pp. 415-422.

[Fu 5] H. Fujimoto, *The defect relations for the derived curves of a holo-morphic curve in* $\mathbf{P}^n(\mathbf{C})$, Tohoku Math. J. **34**, No. 1 (1982), pp. 141-160.

[F-L]   W. Fulton and S. Lang, *Riemann–Roch Algebra*, Springer-Verlag, New York, 1985.

[Ga]   T. Garrity, *A differential geometric criterion for ample vector bundles*, to appear (see also Thesis, Brown University 1986).

[G-K]   S. Goldberg and S. Kobayashi, *On holomorphic bisectional curvature*, J. Diff. Geometry, **1** (1967), pp. 225-233.

[Gr]   H. Grauert, *Mordells Vermutung über rationale Punkte auf algebra-ischen Kurven und Funktionenkörper*, Publ. Math. IHES, **25** (1965), pp. 131-149.

[G-R]   H. Grauert and H. Reckziegel, *Hermitesche Metriken und normale Familien holomorpher Abbildungen*, Math. Z., **89** (1965), pp. 108-125.

[Gr 1] M. Green, *Holomorphic maps into complex projective space omitting hyperplanes*, Trans. Amer. Math. Soc., **169** (1972), pp. 89-103.

[Gr 2] M. Green, *Some Picard theorems for holomorphic maps to algebraic varieties*, Amer. J. Math., **97** (1975), pp. 43-75.

[Gr 3] M. Green, *The hyperbolicity of the complement of* $2n + 1$ *hyperplanes in general position in* P_n, *and related results*, Proc. Amer. Math. Soc., **66**, No. 1 (1977), pp. 109-113.

[Gr 4] M. Green, *Holomorphic maps to complex tori*, Amer. J. Math., **100** (1978), pp. 615-620.

[Gr 5] M. Green, *Some examples and counterexamples in value distribution theory for several variables*, Compositio Math., **30**, No. 3 (1975), pp. 317-322.

[Gr 6] M. GREEN, *The complement of the dual of a plane curve and some new hyperbolic manifolds*, Value-Distribution Theory, Part A, Marcel Dekker, New York, 1974, pp. 119-132.

[G-G] M. GREEN and P. GRIFFITHS, *Two applications of algebraic geometry to entire holomorphic mappings*, The Chern Symposium 1979, Proc. Internal. Sympos., Berkeley, CA., 1979, Springer-Verlag, New York, 1980, pp. 41-74.

[Gri 0] P. GRIFFITHS, *Hermitian differential geometry and the theory of positive and ample holomorphic vector bundles*, J. Math. Mech., **14**, No. 1 (1965), pp. 117-140.

[Gri 1] P. GRIFFITHS, *Hermitian differential geometry, Chern classes and positive vector bundles*, Global Analysis in honor of Kodaira, Edited by D. C. Spencer and S. Iyanaga, Princeton University Press, Princeton, NJ, 1969, pp. 185-252.

[Gri 2] P. GRIFFITHS, *Holomorphic mappings into canonical algebraic varieties*, Ann. of Math., **98**, No. 2 (1971), pp. 439-458.

[Gri 3] P. GRIFFITHS, *Entire holomorphic mappings in one and several complex variables*, Ann. of Math. Studies, Vol 85, Princeton University Press, Princeton, NJ. 1976.

[Gro] A. GROTHENDIECK, *Sur la classification des fibrés holomorphes sur la sphère de Riemann*, Amer. J. Math., **79** (1957), pp. 121-138.

[Gu-R] R. GUNNING and H. ROSSI, *Analytic Functions of Several Complex Variables*, Prentice-Hall, Englewood Cliffs, NJ, 1965.

[Ha 1] R. HARTSHORNE, *Algebraic Geometry*, Springer-Verlag, New York, 1977.

[Ha 2] R. HARTSHORNE, *Ample vector bundles*, Publ. Math. IHES, **29** (1966), pp. 63-94.

[Ha 3] R. HARTSHORNE, *Ample subvarieties of algebraic varieties*, Springer Lecture Notes, Vol. 156, Springer-Verlag, Berlin, 1970.

[H-K] R. HARVEY and A. W. KNAPP, *Positive (p, p)-forms, Wirtinger's inequality, and currents*, Value-Distribution Theory, Part A, Proc. Tulane Univ. Program on Value-Distribution Theory, in *Complex Analysis and Related Topics in Differential Geometry*, Marcel Dekker, New York, 1974, pp. 43-62.

[H-W] W. HUREWICZ and H. WALLMAN, *Dimension Theory*, Princeton University Press, Princeton, NJ, 1948.

[Ki 1] P. KIERNAN, *On the relations between taut, tight, and hyperbolic manifolds*, Bull. Amer. Math. Soc., **76** (1970), pp. 49-51.

[Ki 2] P. KIERNAN, *Hyperbolically imbedded spaces and the big Picard theorem*, Math. Ann., **204** (1973), pp. 203-209.

[Ki 3] P. KIERNAN, *Extensions of holomorphic maps*, Trans. Amer. Math. Soc., **172** (1972), pp. 347-355.

[K-K 1] P. KIERNAN and S. KOBAYASHI, *Satake compactification and the great Picard theorem*, J. Math. Soc. Japan, **23** (1972), pp. 340-350.

[K-K 2] P. KIERNAN and S. KOBAYASHI, *Holomorphic mappings into projective space with lacunary hyperplanes*, Nagoya Math. J., **50** (1973), pp. 199-216.

[Kl] S. KLEIMAN, *Toward a numerical theory of ampleness*, Ann. of Math., **84** (1966), pp. 293-344.

[Ko 1]   S. KOBAYASHI, *Volume elements, holomorphic mappings and the Schwarz lemma*, Proc. Symp. Pure Math., **11** (1968), pp. 253–260, in *Entire Functions and Related Analysis*, American Mathematical Society, Providence, RI.

[Ko 2]   S. KOBAYASHI, *Distance, holomorphic mappings and the Schwarz lemma*, J. Math. Soc. Japan, **19** (1967), pp. 481–485.

[Ko 3]   S. KOBAYASHI, *Invariant distances on complex manifolds and holomorphic mappings*, J. Math. Soc. Japan, **19** (1967), pp. 460–480.

[Ko 4]   S. KOBAYASHI, *Hyperbolic Manifolds and Holomorphic Mappings*, Marcel Dekker, New York, 1970.

[Ko 5]   S. KOBAYASHI, *Negative vector bundles and complex Finsler structures*, Nagoya Math. J., **57** (1975), pp. 153–166.

[Ko 6]   S. KOBAYASHI, *Intrinsic distances, measures and geometric function theory*, Bull. Amer. Math. Soc., **82** (1976), pp. 357–416.

[K–O 1] S. KOBAYASHI and T. OCHIAI, *Satake compacifications and the great Picard theorem*, J. Math. Soc. Japan, **23** (1971), pp. 340–350.

[K–O 2] S. KOBAYASHI and T. OCHIAI, *On complex manifolds with positive tangent bundles*, J. Math. Soc. Japan, **22** (1970), pp. 499–525.

[K–O 3] S. KOBAYASHI and T. OCHIAI, *Meromorphic mappings onto compact complex spaces of general type*, Invent. Math., **31** (1975), pp. 7–16.

[K–O 4] S. KOBAYASHI and T. OCHIAI, *Mappings into compact complex manifolds with negative first Chern class*, J. Math. Soc. Japan, **23**, No. 1 (1971), pp. 137–148.

[Kw]   M. KWACK, *Generalizations of the big Picard theorem*, Ann. of Math., **90**, No. 2 (1969), pp. 9–22.

[La 0]   S. LANG, *Integral points on curves*, Publ. Math. IHES, (1960), pp. 27–43.

[La 1]   S. LANG, *Introduction to Transcendental Numbers*, Addison-Wesley, Reading, MA, 1966.

[La 2]   S. LANG, *Higher dimensional diophantine problems*, Bull. Amer. Math. Soc., **80**, No. 5 (1974), pp. 779–787.

[La 3]   S. LANG, *Hyperbolic and diophantine analysis*, Bull. Amer. Math. Soc., **14**, No. 2 (1986), pp. 159–205.

[La 4]   S. LANG, *Differential Manifolds*, Springer-Verlag, New York, 1985.

[La 5]   S. LANG, *Fundamentals of Diophantine Geometry*, Springer-Verlag, New York, 1983.

[Le 1]   P. LELONG, *Fonctions entières (n variables) et fonctions plurisousharmoniques d'ordre fini dans* $\mathbf{C}^n$, J. Analyse Math., **12** (1964), pp. 365–407.

[Le 2]   P. LELONG, *Plurisubharmonic Functions and Positive Differential Forms*, Gordon and Breach, New York, 1969.

[Ma]   J. MANIN, *Rational points of algebraic curves over function fields*, Izvestia Akad. Nauk, **27** (1963) pp. 1395–1440.

[M–M]   S. MORI and S. MUKAI, *The uniruledness of the moduli space of curves of genus 11*, Algebraic Geometry Conference, Tokyo–Kyoto, 1982, Springer Lecture Notes, Vol. 1016, Springer-Verlag, New York, pp. 334–353.

[Na]   S. NAKANO, *On complex analytic vector bundles*, J. Math. Soc. Japan, **7** (1955), pp. 1–12.

[Ne 1]   R. NEVANLINNA, *Zur Theorie der meromorphen Funktionen*, Acta Math., **46** (1925), pp. 1-99.

[Ne 2]   R. NEVANLINNA, *Einige Eindeutgkeitssätze in der Theorie meromorphen Funktionen*, Acta Math., **48** (1926), pp. 367-391.

[Ne 3]   R. NEVANLINNA, *Analytic Functions*, Springer-Verlag, New York, 1970. Translated and revised from *Eindeutige analytische Funktionen*, 2nd ed., Springer-Verlag, New York, 1953.

[No 1]   J. NOGUCHI, *A higher dimensional analogue of Mordell's conjecture over function fields*, Math. Ann., **258** (1981), pp. 207-212.

[No 2]   J. NOGUCHI, *Logarithmic jet spaces and extensions of de Franchis' theorem*, to appear.

[No 3]   J. NOGUCHI, *Hyperbolic fiber spaces and Mordell's conjecture over function fields*, Publ. Research Institute Math Sciences Kyoto University **21**, No. 1 (1985), pp. 27-46.

[No 4]   J. NOGUCHI, *Meromorphic mappings of a covering space over C^m into a projective variety and defect relations*, Hiroshima Math. J., **6**, No. 2 (1976), pp. 265-280.

[No 5]   J. NOGUCHI, *Holomorphic curves in algebraic varieties*, Hiroshima Math. J., **7**, No. 3 (1977), pp. 833-853.

[No 6]   J. NOGUCHI, *Supplement to "Holomorphic curves in algebraic varieties"*, Hiroshima Math. J., **10**, No. 1 (1980), pp. 229-231.

[No 7]   J. NOGUCHI, *Meromorphic mappings into a compact complex space*, Hiroshima Math. J., **7**, No. 2 (1977), pp. 411-425,

[No 8]   J. NOGUCHI, *Lemma on logarithmic derivatives and holomorphic curves in algebraic varieties*, Nagoya Math. J., **83** (1981), pp. 213-233.

[No 9]   J. NOGUCHI, *A relation between order and defects of meromorphic mappings of C^n into $P^N(C)$*, Nagoya Math. J., **59** (1975), pp. 97-106.

[No 10] J. NOGUCHI, *Holomorphic mappings into closed Riemann surfaces*, Hiroshima Math. J., **6**, No. 2 (1976), pp. 281-291.

[No 11] J. NOGUCHI, *On the value distribution of meromorphic mappings of covering spaces over C^m into algebraic varieties*, J. Math. Society Japan, **37**, No. 2 (1985), pp. 295-313.

[O'N]   B. O'NEILL, *Isotropic and Kähler immersions*, Canadian Math. J., **17** (1965), pp. 907-915.

[Re]   H. RECKZIEGEL, *Hyperbolische Räume und normale Familien holomorpher Abbildungen*, Dissertation, Göttingen, 1967.

[Ri]   D. RIEBESEHL, *Hyperbolische Komplexe Raüme und die Vermutung von Mordell*, Math. Ann., **257** (1981), pp. 99-110.

[Ro 1]   H. ROYDEN, *Remarks on the Kobayashi metric*, Proc. Maryland Conference on Several Complex Variables, Springer Lecture Notes, Vol. 185, Springer-Verlag, Berlin, 1971.

[Ro 2]   H. ROYDEN, *The extension of regular holomorphic maps*, Amer. Math. Soc., **43**, No. 2 (1974), pp. 306-310.

[Ro 3]   H. ROYDEN, *Automorphisms and isometries of Teichmüller space*, Advances in the Theory of Riemann Surfaces, Princeton University Press, Princeton, NJ, 1971.

[Si]   C. L. SIEGEL, *Über einege Anwendungen Diophantischer Approximationen*, Abh. Preuss. Akad. Wiss, Phys. Math. Kl., (1929), pp. 41-69.

[Siu]   Y-T SIU, *Every Stein subvariety admits a Stein neighborhood*, Invent.
           Math., **38** (1976), pp. 89–100.

[Sm]   K. T. SMITH, *Primer of Modern Analysis*, Springer-Verlag, New York,
           1983. Especially Chapter 15.

[Stoll 1] W. STOLL, *About entire and meromorphic functions of exponential type*,
           Proc. Symp. Pure Math., **11** (1968), pp. 392–430.

[Stoll 2] W. STOLL, *The growth of the area of a transcendental analytic set*, (I
           and II), Math. Ann., **156** (1964), pp. 47–98; and **156** (1964), pp.
           144–170.

[Stoll 3] W. STOLL, *Value distribution of holomorphic maps into compact complex
           manifolds*, Springer Lecture Notes, Vol. 135, Springer-Verlag, Berlin,
           1970.

[Stoll 4] W. STOLL, *Value Distribution Theory*, Part B, Marcel Dekker, New
           York, 1974.

[Stoll 5] W. STOLL, *Value distribution theory for meromorphic maps*, Vieweg, As-
           pects of Mathematics, 1985.

[Stoll 6] W. STOLL, *Value distribution on parabolic spaces*, Springer Lecture
           Notes, Vol. 600, Springer-Verlag, Berlin, 1973.

[Stolz]   G. STOLZENBERG, *Volumes, limits, and extensions of analytic varieties*,
           Springer Lecture Notes, Vol. 19, Springer-Verlag, Berlin, 1966.

[Ur]   T. URATA, *The hyperbolicity of complex analytic spaces*, Tull. Aichi
           University of Education, **XXXI** (1982), pp. 65–75.

[Vo]   P. VOJTA, *Diophantine approximation and value distribution theory*,
           Springer Lecture Notes No. 1239, 1987.

[Wu 1]   H. WU, *The equidistribution theory of holomorphic curves*, Ann. of
           Math. Studies, Vol. 64, Princeton University Press, Princeton, NJ,
           1970.

[Wu 2]   H. WU, *A remark on holomorphic sectional curvature*, Indiana Univ.
           Math. J., **22**, No. 11 (1973), pp. 1103–1108.

[Yau]   S. T. YAU, *Intrinsic measures on compact complex manifolds*, Math.
           Ann., **212** (1975), pp. 317–329.

Index

A

Ahlfors–Schwarz lemma   102, 107
Ahlfors–Shimizu   217
Ample   82, 118, 150, 157
Ascoli's theorem   28, 73
Atlas for length function   9
Ax theorem   85

B

Ball   3
Barth's theorem   19
Barton theorem   150
Bianchi identity   131
Bloch   195, 198, 224
Bloch–Cartan estimate   236
Blow down   81
Borel's lemma   177, 228
Borel's theorem   186, 233
Brody–Green   209
Brody hyperbolic   67, 78, 196, 198
Brody hyperbolic modulo S 78
Brody's reparametrization   69
Brody's theorem   66, 68, 223

C

C*-convergent   233
Canonical bundle   99
Canonical product   160, 173
Canonical variety   119

Carlson–Griffiths   114
Cartan height   202
Cartan–Nevanlinna height   213
Cartan's conjecture   244
Cartan's theorems   79, 169, 195, 220, 224
Cartier divisor   75
Characteristic function   167
Chern form   97, 113, 153
Closed ball   3
Complete hyperbolic   24, 75, 77, 198
Complete hyperbolic modulo S 37
Complex connection   133
Complex space   7
Complex torus   83
Connection   124, 133
Contraction   139
Counting function   164, 213
Coverings   23
Curvature   100, 127, 137

D

Degree   179, 213
Derived fraction   251
Diagonals   78, 195
Disc   3
Distance   3
Distance decreasing   13, 16
Distance function   4
Divisor   179
Duality   149

E

Einstein–Kähler   100
Einsteinian   100
Equicontinuous   29
Euclidean form   98
Euclidean measure   49
Exceptional set   123

F

Fermat variety   204
First Main Theorem   168, 214
Frame   124, 137
Fubini–Study   98, 218
Fujimoto   196, 215

G

Garrity   154
Gauss curvature   100
Gauss function   145
General position   78, 194, 220
General type   119
Geodesic   13
Grauert's criterion   83
Green theorems   75, 83, 196, 198, 199,
    205
Green's example   79
Griffiths function   100, 106, 107
Griffiths positive   139

H

Hausdorff measure   53
Height   167, 201, 213, 241
Hermitian bundle   132
Hermitian connection   133
Hermitian length function   103
Hermitian manifold   102
Holomorphic   7
Holomorphic bundle   132
Holomorphic frame   137
Holomorphic sectional curvature   140
Hyperbolic   19, 23, 27, 68, 75, 81, 83,
    85, 106, 157, 209
Hyperbolic form   100, 107, 114
Hyperbolic measure   43
Hyperbolic metric or norm   11
Hyperbolic modulo a subset   36, 78
Hyperbolically imbedded   32, 40, 75,
    77, 85, 198, 245
Hyperbolically imbedded modulo a
    subset   37, 78, 79, 245

J

Jensen formula   162, 241

K

Kiernan–Kobayashi–Kwack (K^3)   58
Kleiman theorem   150
Kobayashi chain   15
Kobayashi hyperbolic   19
Kobayashi length   15
Kobayashi–Ochiai   122
Kobayashi path   15
Kobayashi semi distance   15
Kobayashi sum   15
Kodaira   122
Kwack's theorem   40, 58

L

Lelong's theorem   51
Length   3, 10, 11
Length function   1, 3, 9, 103
Lifting   194
Local imbeddings   9
Locally compact (function spaces)   29
Locally complete hyperbolically   35
Locally relatively compact   30
Logarithmic derivative   172, 187, 189

M

Measure decreasing   110
Measure hyperbolic   110, 122
Metric   94
Montel   230

N

Nakano positive   140
Nevanlinna   167, 171, 181
Noguchi theorems   56, 61, 74
Norm   6
Normal crossings   58, 61, 114, 116,
    181, 214
Normal family   73, 226, 230, 233
Normalized hyperbolic form   100, 107
Numerical equivalence   147

O

Open ball   3

P

Path 4, 10
Picard's theorem 196
Poisson 159
Poisson–Jensen 161, 242
Positive 2
Positive form 97, 101
Positive metric 97
Product formula 163
Projective bundle 81, 148
Projective space 97
Proper map 72
Proximity function 167, 213, 217
Pseudo ample 119
Pseudo Brody hyperbolic 123
Pseudo canonical 119
Pseudo Kobayashi hyperbolic 123
Pseudo volume form 108, 122

R

r-degree 179
Ramification 180
Regular point 8
Relatively locally compact 29, 30, 35
Relatively locally compact modulo S 37, 79
Reparametrization lemma 69
Resolution 8
Ricci form 99
Ricci function 140
Ricci product 139
Ricci tensor 136
Riemannian metric 49
Royden function 88

S

Schwarz–Pick lemma 13
Second Main Theorem 181, 214, 220
Semi length function 2

Simple divisor 181
Simple normal crossings 58, 114, 181, 214
Singular point 8
Sphere 3
Strongly hyperbolic form 105, 113, 114
Subharmonic 46

T

Tautological line bundle 97
Tautological metric 97
Tensors 137
Totaro 122

U

Units 185
Universal quotient bundle 148
Universal subbundle 151
Upper semi continuous 2
Urata 81

V

Valuation 163
Very ample 119
Very canonical 119
Volume form 43, 99, 100

W

Weil function 212
Wirtinger's theorem 50
Wronskian 188, 228
Wu's theorem 144

Z

Zariski open set 79
Zero section 81